HANDBOOK ON ENVIRONMENTAL ASPECTS OF FERTILIZER USE

W0080183

Handbook on Environmental Aspects of Fertilizer Use

N CENTRE D'ÉTUDE DE L'AZOTE (CEA)
Bleicherweg 33
CH–8002 Zürich

P INTERNATIONAL FERTILIZER INDUSTRY
ASSOCIATION (IFA)
28, Rue Marbeuf
F–75008 Paris

K INTERNATIONAL POTASH INSTITUTE (IPI)
P.O. Box 41
CH–3048 Worblaufen-Berne

1983

MARTINUS NIJHOFF / DR W. JUNK PUBLISHERS
THE HAGUE / BOSTON / LONDON

for the CEA IFA IPI

Distributors

for the United States and Canada

Kluwer Boston, Inc.
190 Old Derby Street
Hingham, MA 02043
USA

for all other countries

Kluwer Academic Publishers Group
Distribution Center
P.O.Box 322
3300 AH Dordrecht
The Netherlands

This volume is listed in the Library of Congress Cataloging in Publication Data

Main entry under title:

Handbook on environmental aspects of fertilizer use.

   Result of a cooperative effort by the CEA,
ISMA, and the IPI through their NPK Working Group on
Environmental Aspects of Fertilizer Use.
   Includes bibliographical references.
   1. Fertilizers--Handbooks, manuals, etc.
2. Fertilizers--Environmental aspects--Handbooks,
manuals, etc. 3. Fertilizers--Europe--Handbooks,
manuals, etc. 4. Fertilizers--Environmental aspects
--Europe--Handbooks, manuals, etc. I. NPK Working
Group on Environmental Aspects of Fertilizer Use.
S633.H28 1983   631.8'1   82-24599
ISBN-13: 978-90-247-2801-5   e-ISBN-13: 978-94-009-6816-5
DOI: 10.1007/978-94-009-6816-5

90-247-2801-0 (this volume)

Martinus Nijhoff / Dr W. Junk Publishers, P.O.Box 566, 2501 CN The Hague, The Netherlands.

Contents

1

Introduction

Chapter A

World population is increasing and more food is needed to support this increase. Nutritional standards over much of the world are at present below the desirable standard — these standards must be raised. Both these requirements entail increasing agricultural production. Food sypply is not the only concern. The products of agriculture include many other necessities: raw materials for industry which contribute to economic welfare. This aspect is very important in parts of the developing world which lack mineral resources.

Chapter B

Agricultural production can be expanded by bringing more land under cultivation or by increasing the yield from the existing cultivated land. In most parts of the world there is little or no reserve of cultivable land; there is competition to use this resource to cater for other, non-food, needs: buildings, roads, factories, amenities.

Agricultural production per unit area of land must be intensified; the stark alternative is starvation or, at the least, an unacceptable degree of malnutrition and economic stagnation.

Chapter G

There are many ways to increase yields: better cultivation, better crops (plant breeding), control of pests and diseases by biological or chemical methods, drainage, irrigation and, most important, higher soil fertility.

Chapters C, G

Improving soil fertility means improving the supply of plant nutrients so that crop growth is not restricted. We should always make the maximum possible use of farm and other residues which will greatly assist in maintaining soil fertility but, if we are to raise the fertility of an area, and in most cases if we are to maintain it, extra nutrients must be brought in from outside. There are two possible ways to do this: (a) by importing food and animal feed, which may involve exploiting some less fortunate, and less percipient, part of the world; this is implicit in some of the so-called "alternative" methods of farming, and (b) by using fertilizers which make use only of the atmosphere (for nitrogen), mineral deposits and fossil energy.

Fertilizer is essential if we are to survive

Chapters D, E, F

Though it is beyond question that fertilizer is an essential feature of modern productive farming and that it must be used to improve less advanced farming systems, criticism of its use is voiced from time to time. Such criticism is invalid alongside the obvious benefits from using fertilizer. Nevertheless, it may be helpful to point out that, when examined objectively, these criticisms are found to be groundless. Far from adversely affecting the quality of crops and animal products, fertilizer, when properly used, improves it.

Chapters F, H

Some critics have attempted to implicate fertilizer in pollution of the environment. These criticisms are in general ill-founded.

Chapters C, G, J, K

Of course, it is possible to misuse fertilizer, as any other product and some criticisms may stem from isolated instances of misuse. When fertilizer is properly used, the effects are entirely beneficial whether our concern is with soil fertility, crop yield, crop quality, health, the environment or the quality of human life. And fertilizers help improve the productive use of solar energy.

General note

The aim of this handbook is to provide Public Relations Officers and others within the Fertilizer Industry with information which can be made use of in dealing with enquiries from the Press (including radio, television, etc.) and the Public in general. The text is the result of a cooperative effort by the Centre d'Etude de l'Azote, the International Superphosphate and Compound Manufacturers Association (now IFA) and the International Potash Institute through their NPK Working Group on Environmental Aspects of Fertilizer Use. It is *not* intended that the material should be quoted verbatim but rather that it should be interpreted according to the circumstances prevailing in the country or district concerned at the time when an enquiry is made.

Although the information that is given is an accurate representation of the present state of scientific knowledge in this field in Western Europe, the Associations cannot accept liability for any loss, legal action or other difficulties which might be attributed to its use.

Terminology and units

A distinction is drawn in this Handbook between "fertilizers" and "organic manures". The word "fertilizers" refers to commercial fertilizers produced by industry, comprising mainly inorganic, or mineral, fertilizers and urea but also including organic fertilizers which are wastes from industrial processing of agricultural products (e.g. shoddy, bone meal, hoof-and-horn meal, dried blood). The term "organic manures", on the other hand, refers to wastes and residues mainly produced on the farm, such as farmyard manure, slurry, compost and green manure crops for ploughing-in, but it also includes manures made from town wastes, such as sewage sludge and pulverized refuse. Most organic manures are bulky materials containing much carbon and relatively small percentages of plant nutrients, in contrast to fertilizers which in general are much higher in plant nutrient content.

Plant nutrients are expressed, according to the context, as elements (N, P, K, etc.) or as oxides (e.g. P_2O_5, K_2O) or as ions (NH_4, PO_4, etc.).

Metric (SI) units are used throughout, except where otherwise stated.

"Western Europe", unless otherwise stated, means Austria, Belgium, Denmark, Finland, France, the Federal Republic of Germany, Greece, Iceland, the Republic or Ireland, Italy, Luxemburg, Malta, the Netherlands, Norway, Portugal, Spain, Sweden, Switzerland, the United Kingdom of Great Britain and N. Ireland, and Yugoslavia.

"EEC", unless otherwise stated, means the European Economic Community of "The Ten" (Belgium, France, Federal Republic of Germany, Greece, Italy, Luxemburg, the Netherlands, Denmark, the Republic of Ireland, and the United Kingdom).

"Developed Regions" and "Developing Regions" are as defined by FAO.

4

A Food, fibre and fertilizers

A 1   Food and fibre needs of the world

The greatest problem facing the World today was put in a nutshell by the
Director General of FAO when he opened the Conference "Agriculture
toward 2000" in 1979. In the past twenty years world population had
increased from 3 billion to approximately 4½; by the end of the century it
would be more than 6 billion. Even to feed the people at present average
standards, and we know that these standards are far from adequate in many
parts of the world, would involve growing one and a half time as much food
as we now produce. Where, and how, is this food to be grown?

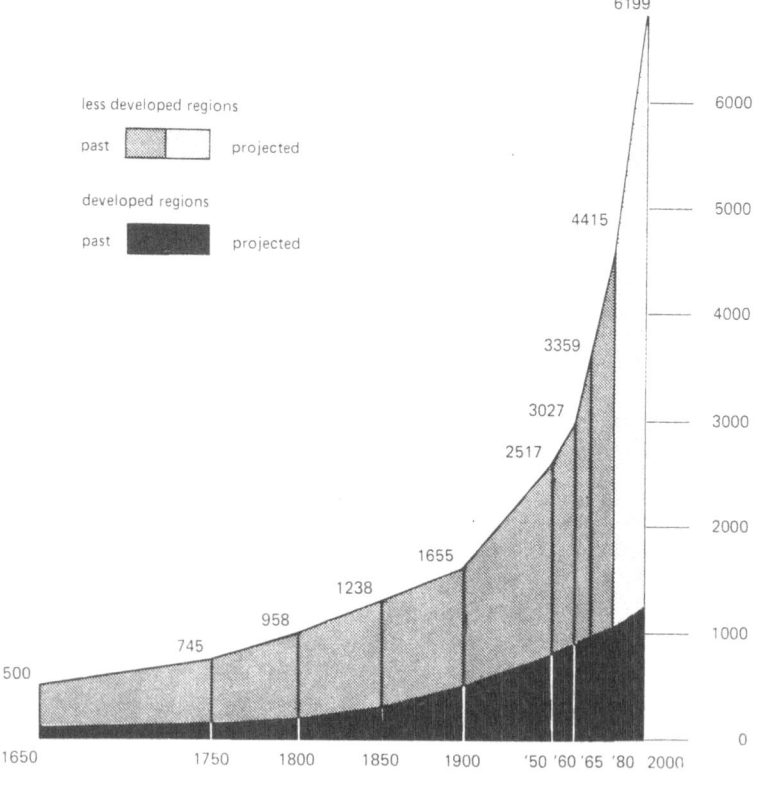

Figure A 1.1 World population (millions) since 1650, with the projection to 2,000 A.D.

Table A 1.1. *Present average daily diets and projected targets for the developing countries in 2000 A.D.*

Item	Developed countries	Developing countries	
		1979	target for 2000
Energy, kjoules per head*			
Estimated requirement	11,000	9,700	–
	(2,600)	(2,295)	(–)
Present diet	14,050	9,250	11,200
	(3,315)	(2,180)	(2,637)
Percentage undernourished	0	23	7
Protein, g per head			
Total protein	90	58	76
Animal protein	45	10	20
Fat, g per head	110	38	54

* between parentheses kcal.

Food supply is not the only problem. The raw materials of many industries are non-food or cash crops, e.g. rubber, fibres, tobacco or non-essentials like tea, coffee, cacao, spices which offer the developing regions the opportunity to take part in world trade and are important sources of foreign exchange. If present standards of economic well-being are to be improved, or even maintained, the output of these materials must also be increased. This places a further demand on world agriculture.

Agricultural production can only be increased either by bringing more land under cultivation or by increasing crop yield per unit area. Cultivable land is a finite resource and FAO estimates show that opening up more land to farming could make possible only one quarter of the increase in food and fibre production which is needed; 74% of the increase must come from an increase in crop yield (Table A 1.2).

Table A 1.2. *Expected contributions to the required increase in agricultural output 1980–2000.*

Region	From increased area under cultivation, %	From increased yield per unit area, %
90 Developing Countries	26	74
Africa	27	73
Far East	10	90
Latin America	55	45
Near East	6	94

6

This is the average situation; in heavily populated parts of the world there is scarcely any reserve of cultivable land.

Crop yield per hectare can be increased by: better cultural methods, better drainage and irrigation, introducing improved crop varieties, effective protection against pest and diseases and by improving soil fertility. The various factors of improvement are inter-related; for example, the heavier crops made possible by using better varieties involve a heavier demand on plant nutrients from the soil. If the crop's food demand is not covered, it cannot yield at its full potential and the benefit offered by the improved crop will be lost. Fertilizer is an essential component of improving soil fertility. Because of the complex way in which the crop production factors operate, it is difficult to estimate exactly how much of the yield increases achieved in the past has been due to the use of fertilizers and how much to other factors, but FAO estimates that fertilizers have been responsible, directly and indirectly, for over one half (56%) of the increase in crop production brought about in the past thirty years.

A 2   Improving soil fertility

The problem of improving soil fertility is dealt with in more detail in Section G. In summary, the recycling of agricultural and other residues and waste products can contribute to maintaining soil fertility but it cannot improve it. Raising soil fertility involves adding nutrients from outside the system. This can be done only by using fertilizers. At the present time total world usage of fertilizers amounts to about 119 million t of plant nutrients, $(N + P_2O_5 + K_2O)$, over 85% of which is used by developed countries (Table A 2.1). FAO estimates that the developing countries will need to increase fertilizer usage by five times if they are to be able to feed their people in the year 2000.

Table A 2.1. *Fertilizer usage in developing countries and projected requirements to 2000 in million t $N + P_2O_5 + K_2O$.*

Region	1965	1975	1980	1985	1990	2000
90 Developing Countries	4.1	13.8	19.0	28.4	47.1	92.9
Africa	0.3	0.8	1.0	1.5	2.3	5.3
Far East	1.7	6.5	9.8	15.2	28.9	57.6
Latin America	1.5	4.4	5.4	7.4	10.1	18.6
Near East	0.6	2.1	2.8	4.0	5.8	11.4

The working group of FAO, UNIDO and Worldbank have made estimates of fertilizer requirement in more detail but over a shorter period. They estimate that in ten years' time the world will need 173 million t of fertilizer nutrients (86 million t N, 49 million t P_2O_5 and 38 million t K_2O).

The competent world authorities agree that there is no way to meet agricultural production targets other than by greatly increased use of fertilizers. An approximate "rule of thumb" is that each ton of balanced plant nutrients (N, P_2O_5, K_2O) used in the developing world could increase cereal production by 10 t.

A 3   The contribution of fertilizers

In developed areas, agricultural production, aided by the inputs provided by
modern technology, has increased more rapidly than has the population. The
result has been that the peoples of these areas are now better fed from their
own farming land than they were at the beginning of the century despite the
fact that the area cultivated has actually decreased.

In the developing countries as a whole the rate of population increase is
higher than in the developed but agricultural production, profiting less from
modern aids, is increasing more slowly and over the past ten years production per head of the population has remained at much the same level or has
even declined in some countries. In some of these countries population is
increasing at 3% per year; the problem of feeding the future population is
immense. While in the developed world food supplies are in excess of the
minimum requirements of the people for an adequate diet, food supplies in
the developing world are insufficient to supply even the minimum need (see
Section A 1).

The two examples following show that has been achieved in a developed
area (Western Europe) and in a developing area (India) and what has been the
contribution of fertilizer to these achievements.

Western Europe

Agricultural output has grown faster than the population. In the 70 years
since 1910 while the population increased by 70%, wheat and sugar output
have trebled, and maize has gone up by eight times. Much more milk and
meat is being produced (Table A 3.1).

Food production has been greatly increased despite a reduction in the area
of cultivated land caused by the demands for housing, industrialization, roads
and airfields. It is estimated that within the EEC the agricultural area shrank
by 8% between 1938 and 1972 while population increased by 31%. However,
home produced food supplies almost doubled so that the net result was an
increase in food production of 46% per head.

Table A 3.1. *Population and food supplies in Western Europe, 1910–1979.*

Year	Population, million	Production, million t				
		Wheat	Maize	Sugar beet	Milk	Meat
1910	232	21.1	3.9	29.0		
1930	256	22.3	4.3	39.3		4.9
1950	297	25.9	2.9	38.1	76.6	5.6
1970	340	44.9	14.2	72.5	127.7	18.8
1979	370	60.3	32.1	99.8	134.7	27.5

Over the same period fertilizer usage has increased many times – by a factor of 5 since 1930 when statistics started (Table A 3.2).

Table A 3.2. *Fertilizer consumption in Western Europe 1930–1978.*

Year	Million t nutrients			
	N	P_2O_5	K_2O	Total
1930	0.7	2.6	1.3	4.6
1950	1.5	2.2	2.2	5.9
1970	6.3	5.2	5.1	16.6
1980	10.0	6.1	5.7	21.8

It may be noted (see also Table A 3.1) that the spectacular increase in food production came about after 1950, coinciding with the steep increase in fertilizer use which followed the second world war. It is estimated that the proportion of agricultural output which may be attributed to fertilizers has increased from about 10% in 1910, when little (mainly phosphate) was used, to about 50% today. While it is true that such production increases would not have been possible without the contribution of the plant breeders by way of improved varieties and without advances in other branches of agricultural technology, it is equally true that the benefits of these other improvements would not have been realised without fertilizers. In the words of Professor Noirfalise, EEC consultant (1974): "Fertilizers account for only about 10% of farming costs, but their effect on productivity is certainly 50%. One cannot therefore conceive of fertilizer use being curtailed or abandoned without serious consequences both to the profitability of farming and to the supply of foodstuffs to the population as a whole at the prices to which they are accustomed and above which any substantial increase would be hard to bear". The UNESCO Courier reported thus in 1971: "Henceforward, half the world population is fed from the additional production of food resulting from the use of fertilizers".

It is interesting to speculate upon what might happen if fertilizers were suddenly no longer available to European farmers. In the first place, food production would decrease by an amount equivalent to the needs of 100 million people whose needs would have to be met by imports, the cost of which, even supposing that this food would be available from other parts of the world, would impose a heavy burden on the economy. In any case, taking the world as a whole, food supply is in deficit; in much of the developing world there is insufficient food available to supply the population with the bare minimum needed for adequate nutrition.

10

India

All are familiar with the problems faced by India in feeding her growing population. In the fifties and sixties famine conditions were frequent and India had to rely heavily on food aid from the developed world. Since those days grain production has increased by 50%; this was achieved almost entire-

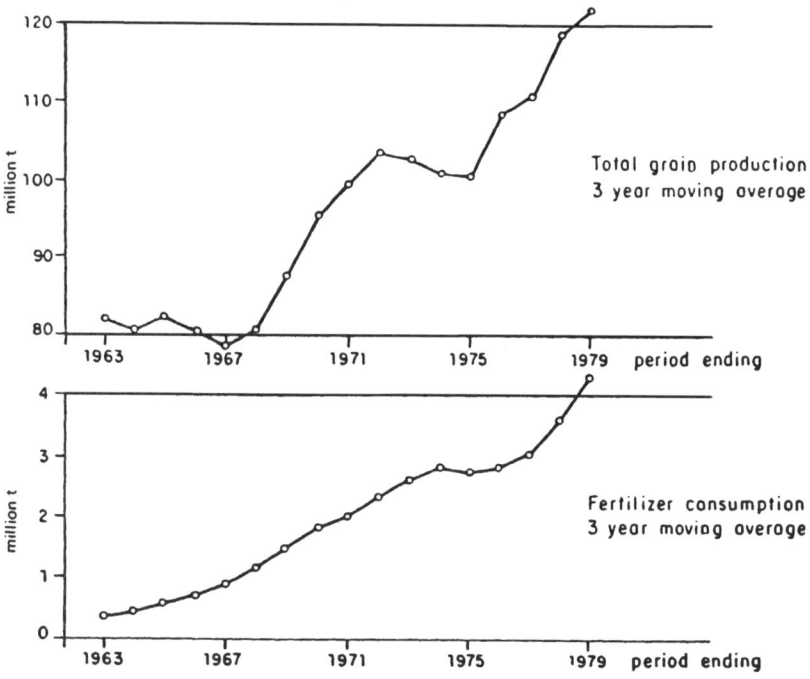

Figure A 3.1 Grain production and fertilizer usage in India 1963–1979.

ly by improving yield per hectare since the area planted was increased by only 5%. It would be idle to pretend that the whole of this increase has been due simply to the increased use of fertilizer. It would not have been possible without the introduction of new, high yielding crop varieties, without better husbandry standards and more irrigation and, most important, without determined action by the Government. But, the increase would not have been possible without fertilizers. Since 1967 the total grain crop has grown from about 80 million t to over 120 million t: over the same time, fertilizer consumption has grown from a little under a million to 4½ million t per year. Wheat production has grown from 10.4 million t in 1966 to 36.5 million t in 1981 and the average yield per hectare increased from 940 to 1585 kg. This large increase in grain production means that India, which formerly imported large tonnages of grain, is now at, or nearing self sufficiency and she has been able to pay her oil bills by exporting wheat to Russia.

References

Arnon, I. (1981) Modernization of agriculture in developing countries. Resources, potentials and problems John Wiley and Sons, Chichester/New York/Brisbane/Toronto. pp. 565.

Buchner, A. and Sturm, H. (1980) Gezielter Düngen. DLG Verlag, Frankfurt/M. pp. 319.

FAO (1980) Fertilizers in agricultural development — pricing policies and subsidies.

FAO (1981) Agriculture: towards 2000.

FAO (up to 1980) Fertilizer yearbook.

FAO (up to 1980) Production Yearbook.

Gervy, R. (1980) The contribution of fertilizers to food production. Phosphorus in agriculture No. 79, 17–36.

Greenland, D.J. (1981) Soil management and soil degradation. J.Soil Sci. 32, 301–322.

Greenwood, D.J. (1981) Fertilizer use and food production: world scene. Fertilizer Research 1, 33–51.

Mahler, H. (1980) Weltbevölkerung. Spektrum der Wissenschaft, November, 75–89.

Skrimshaw, N.S. and Taylor, S. (1980) Welternährung. Spektrum der Wissenschaft, November, 63–73.

Swaminathan, M.S. (1980) Past, present and future trends in tropical agriculture. In: Perspectives in world agriculture. Commonwealth Agric. Bureaux, Farnham Royal, 1–47.

B Social and environmental benefits of fertilizers

B 1 Farming profit and the rural standard of living

Fertilizer is an essential input to modern farming; the average yield increases produced by fertilizer repay several times the cost of buying and applying fertilizer. An example from Denmark (Table B 1.1) shows the increases in production resulting from the use of optimum rates of N fertilizer, with costs and gross returns. No allowance is made for the extra value of the increased protein content of the crops.

Table B 1.1. *Value of increased yield of fodder crops and cost of N fertilizer applied. Danish experiments 1963–1973.*

Crop	Fertilizer N		Yield increase	
	kg per ha	Cost, DKr	Feed units per ha (net)	Value, DKr
Grass	450	900	4900	2450
Barley	100	200	1390	1112
Fodder beet	125	250*	1490	745

* plus farmyard manure

A study of Federal German statistics shows that, as rates of fertilizer usage have increased, so has farm income (Table B 1.2).

Table B 1.2.

Period	Average fertilizer use, kg N+ P_2O_5 + K_2O per ha	Farm income, DM per ha*
1964/65–1966/67	200	890
1976/77–1978/79	252	1090

* mixed farms at 1962/1963 prices

It was estimated that crop increases from fertilizers in the USA in 1960–1964 were worth 2½ times the fertilizer cost – a better ratio than that for other farm inputs.

In livestock farming, using fertilizer to improve grass yield allows stocking rate per hectare to be increased and lessens the need for buying concentrate feeds. Classifying French grassland farms on the basis of fertilizer cost per hectare showed the following general relationship between expenditure on fertilizer and management and investment income (Table B 1.3).

Table B 1.3.

Fertilizer expenses per ha, French Francs (1972)	Management and investment income, French Francs
less than 120	less than 600
120–180	600–1000
180–240	1000–1400
over 240	over 1400

As shown below (Table B 1.4) the use of fertilizer in England greatly increased gross margins per cow and per hectare.

Table B 1.4. *Performance of 106 recorded herds (1979)*

Item	Bottom 25% of farms	Top 25% of farms
Nitrogen applied (kg N per ha)	269	338
Cows per ha	1.85	2.36
Litres milk per ha	9798	14075
Gross margin £ per cow	265	386
Gross margin £ per ha	491	912

Inflation has affected the price of fertilizers and some are inclined to say that they are over-priced. But, in recent years, the prices charged for fertilizers have risen less than those for other industrial products and the ratio fertilizer price to farm produce price is now more favourable than was the case ten years ago.

The examples above from the developed countries make it clear that fertilizer is effective in increasing farm profit. Quite apart from increasing overall profitability this has meant that, because productivity per hectare is raised, many smaller farms have been able to remain in business so permitting the continued existence of vigorous local communities in districts where small farms predominate.

Fertilizers can make a special contribution in the developing countries. Soil fertility is often low and the spectacular results given by fertilizer can be the most important and telling first step in persuading the traditional farmer to adopt new methods. As a rough guide, it is estimated that a ton of plant nutrients ($N + P_2O_5 + K_2O$) will produce an extra 10 t of cereals, enough to feed 40 people at 11,000 kjoules (2600 kcal) a day for a year.

As well as improving total agricultural production so that the growing population of the developing world can be fed, it is as important to improve the quality of life of the rural population. The example below demonstrates that intensifying farming, with the aid of fertilizer, can lift the small farmer right out of the subsistence class and make him a comparatively well-off citizen (Table B 1.5).

Table B 1.5. *The effect of increasing improvement on a small traditional rice farm of 1.9 ha with 1.5 ha growing rice.*

Management	Yield, t per ha	Total farm yield t	Home consumption, t	Marketable surplus, t
1. Traditional subsistence farming	1.5	2.25	1.75	0.5
2. Improved varieties, water and weed control	2.2	3.3	1.75	1.55
3. As (2) plus pesticides and fertilizer	3.4	5.1	1.75	3.35
4. As (3) with additional fertilizer and double cropping	8.5	12.75	2.0	10.75

More surplus rice is available for sale to swell the farmer's cash income: alternatively, because a smaller area is needed to provide subsistence, the farmer can diversify, possibly growing a high-value cash crop or a greater variety of food crops which will benefit the health and well-being of his family.

B 2   Fertilizers and the competition for land

Until fairly recently, as the population gradually increased and the demand for food rose, farmers were able to grow the extra needed simply by bringing more land under cultivation. Because a limit must eventually be imposed by the availability of suitable land, it was thought that, in the long run, world famine would be inevitable. The introduction of fertilizers changed all that by making it possible to grow larger crops, and the increase in yield has been such that today the total cultivated area is smaller than it used to be, yet the much larger population is better fed than ever before. Fertilizers have made it possible to concentrate farming on the most productive land and to free other areas to accommodate our needs for housing, industry, communications, forest products etc., and, scarcely less important, for recreation.

In France, for example, the population has increased by 40% since 1862 but the area under tillage has dropped by 35%, enabling more land to be put under grass and forest (Table B 2.1).

Table B 2.1. *Population and land-use in France.*

	Population, million	Million hectares		
		Tillage	Grassland	Forest
1862	35	26	6.8	9.3
1937	40	20	11.8	10.7
1979	53	17	12.9	14.5

Some years ago, the U.S. Department of Agriculture calculated that crop needs for 1980 could be met either from 180 million ha cropland using 7.9 million t fertilizer or from as little as 120 million using 24.2 million t — a saving of 3.7 ha per tonne of fertilizer used.

Land is a scarce resource and has to provide for all the competing demands of modern society. By making land more productive, fertilizers greatly increase the value of this resource making it possible to cater for all our needs even including recreation and amenity.

B 3   Fertilizers and the environment

Fertilizer, properly used, has beneficial effects on the environment. In the first place, by reducing pressure on the land it makes it possible to avoid cultivating steeper, less fertile areas, releasing them for other purposes: forestry, recreational areas, nature reserves, thus greatly improving the quality of human life and enabling us to conserve some of the natural vegetation with its native flora and fauna. A good plant population, actively photosynthesising (producing oxygen) and with a well-developed root system, reduces the risk of erosion and cleans both the air we breathe and the water we drink.

Erosion Fertilizers have sometimes been wrongly blamed for causing land deterioration, particularly increasing erosion. The true situation is the reverse. When fertilizer is used, plant growth is improved, crops establish more rapidly giving a complete ground cover. The crop cover protects the soil against the erosive effects of rain and wind; it improves soil permeability to water so that surface run-off is greatly reduced. The crop cover reduces wind speed and makes the micro-climate more equable. Plant roots bind the soil together and the greater residues from larger crops increase the organic matter content of the soil, thus improving soil structure.

A practical example from the USA shows that fertilizer reduced run-off from a maize field by one third and actual soil loss through erosion by two fifths. Soil loss from land under a rotation of maize, wheat, grass and legumes was reduced by a half.

Water Water supplies may become polluted by run-off of surface water or by leaching into the subsoil. A vigorous and dense plant cover, whether of a crop, grass, or forest, minimises surface water movement and reduces leaching of nutrients, since the downward moving nutrients are taken up by the vigorous root system penetrating the soil. Thus anything which improves plant growth also reduces pollution. Some experimental results from Germany make this clear (Table B 3.1).

The loss of nitrogen by leaching on a sandy soil was also measured in lysimeter experiments at the Limburgerhof Experiment Station, FR Germany. There was no leaching of N when up to 80 kg per ha N was applied; when 160 kg per ha N was applied, 3% of the fertilizer N was found at 1 m depth and even when 240 kg per ha was given, only 7% was washed down to 1 m.

Table B 3.1. *Crop removals and leaching losses of nitrogen as affected by cropping.*

Crop	Nitrogen, kg per ha	
	Removed by crop	Lost by leaching
Grass	190	10
Vines	80	80
Fallow	–	160

In comparison with farm and other wastes used as manures, using fertilizers gives better control over leaching because the amounts of readily available nutrients applied can be more or less exactly equated with the crop's needs and application can be timed to coincide the crop's maximum demands (see also Chapter C).

References

FAO (1965) Soil erosion by water. Agric. Development Paper No. 81, 63–70.

FAO (1972) Effects of intensive fertilizer use on the human environment. Soils Bull. No. 16, pp. 360.

FAO (1980) Fertilizers in agricultural development – pricing policies and subsidies.

Habjorg, A. (1973) Air pollution and vegetation. 11. Effects of fertilization on growth and development of twenty woody plants grown on industrial areas. Sci. Rep. Agric. Univ. Norway 52.

Hernando Fernandez. V. (Ed.) (1974) Fertilizers, crop quality and economy. Proc. study week on use of fertilizers and their effect on crop growth, with emphasis on crop quality and economy. Elsevier Scientific Publ. Co., Amsterdam/ Oxford/New York. pp. 1415.

Holmes, J.C. (Ed.) (1978) Technology for increasing food production. Proc. 2nd FAO/ SIDA Seminar on field food crops in Africa and the Near East, Lahore, Pakistan, pp. 695.

Proceedings of a water seminar. United Nations Economic Commission for Europe, Vienna, 15–20 October 1973.

Royal Society London (1977) Agricultural efficiency. A Royal Society discussion on: The management of inputs for yet greater agricultural yield and efficiency. 17–18 November 1976. University Press Cambridge, pp. 301.

C Alternatives to fertilizers

C 1 Legumes versus fertilizer nitrogen

Legumes play an important part in world agriculture in two ways: First they contribute to food supply directly (grain legumes) and indirectly through the animal (forage crops and clover in mixed grassland swards). Second, through the process of nitrogen fixation they add nitrogen to the soil. Though grain legumes are an important source of protein for the human diet, these crops are, in comparison to cereals, relatively unproductive and it would be impossible to cover the world's protein needs by growing them in place of the cereals and other non-leguminous crops which cover most of the world cropped area. It is also unfortunately true that grassland, certainly in temperate zones, can rarely sustain enough clover to be as productive as grass which receives N fertilizer. The nitrogen contributed to the soil by these crops offers a source of economy but it is not sufficient to support the high yields of arable crops which can now be grown using N fertilizer. The nitrogen provided by legumes is obtained without cash expenditure, but energy is expended in the fixation process in the plant. In a properly nourished crop, 400 kg of carbohydrate is used up in fixing every 100 kg N from the atmosphere.

It might be thought that using legumes to supply N would be less damaging to the environment than using fertilizer; the reverse is often the case. When the plants die at the end of the growing season, there is no crop growth to take up the nitrate which is released from the nodules and plant residues, so there is a great danger of leaching by the winter rains. Leaching losses as high as 130 kg N per ha from soil under red clover have been reported in England and there are many other references to this kind of loss. In such practical situations using legumes may be environmentally less satisfactory than using fertilizers which allow the N supply to be adjusted to the crop's needs. This is important in water catchments.

References

Anonymous (1981) Legumes and fertilizers in grassland systems. Winter meeting, British Grassland Society.

Beringer, J.A. and Day, J.M. (1981) The role of biological nitrogen fixation in UK agriculture. The Fertiliser Society London, Proc. No. 205. pp. 16.

Isermann, K. (1980) Quantitative und qualitative Aspekte des Energiebedarfes bei Ernährung der Kulturpflanzen mit technisch, atmosphärisch und biologisch fixiertem Stickstoff. Vortrag Botanikertag, Bochum, 26–31 May, pp. 18.

Larue, T.A. and Patterson, T.G. (1981) How much nitrogen do legumes fix. Adv. in Agronomy, 34, 15–38.

Low, A.J. and Armitage, E.R. (1970) The composition of the leachate through cropped and uncropped soils in lysimeters compared with that of the rain. Plant and Soil 38, 393–411.

Van Mil, M. (1981) Actual and potential nitrogen fixation in pea and field bean as affected by combined nitrogen. Thesis, Agric. Univ. Wageningen, the Netherlands. pp. 128.

C 2   Farm manures, composted town refuse and sewage sludge

In the interests of conserving soil fertility, all crop and animal residues should be returned to the land. In any case, returning these residues to the land offers the only rational means of disposal and prevention of pollution. It is unfortunate, however, that even when all residues are returned there can be no addition to soil fertility unless the farming system relies heavily on imported feedstuffs. All the farmyard manure produced in the EEC is already fully utilised but this supplies only 4.5 million of the 12 million t of N which are needed by crops, leaving 7.5 million t N to be supplied by N-fertilizers. Though the manure contains significant amounts of phosphate and potash, the EEC still needs to apply 4.7 million t P_2O_5 and 4.5 million t K_2O as fertilizer to support present-day crops. Incidentally, one advantage of using fertilizer is to increase the production of animal manures by enabling stocking rates to be increased.

Other possible sources of nutrients for crops are sewage and town refuse. Using these materials presents certain hazards (see below) but, this apart, the following calculations show that these materials could only supply quite limited amounts of plant nutrients. All the sewage sludge now available in a year in the EEC (12 g N per head per day from 297 million people) contains only 1.3 million t N. When treated in sewage plants, three quarters of this is lost by denitrification leaving 325.000 t of which about 1/5 would be immediately available to crops. So far as town refuse is concerned, average production is about 2 cubic metres or 400 kg per head per year of which half is compostable organic material containing 0.3% N. The total for the EEC would be about 180,000 t of N per year and of this only 1/5th would be of use to crops in the year of application with a further 40% in later years. Thus the total nitrogen available from both these urban wastes would only amount to some 300,000 t of N, a small fraction of that needed by crops grown in the EEC.

A serious disadvantage of the waste materials is that the low concentration of nutrients (about 1/40 of that in the average fertilizer) means that the cost of transport and spreading per unit of nutrient is high and prohibitively so when the distance from source is large.

There are certain hazards in using these waste materials in agriculture. Sewage sludge, especially from industrial areas, may contain considerable amounts of heavy metals. If this is so, its use must be restricted so as to prevent heavy metal accumulating to levels which are toxic to plants and damaging to the health of animals. As an example, in 1980, analysis of 114 sludges from the Rhineland-Palatinate indicated that permissible levels of heavy metal content were frequently exceeded (Table C 2.1).

Table C 2.1

Heavy metal	mg per kg dry matter (DM) in sludge		Upper level permitted, mg per kg DM
Lead	20.0–2,000	(ϕ 297)	500
Cadmium	1.1– 46.5	(ϕ 3.9)	20
Chromium	8.0–4,520	(ϕ 654)	250
Copper	29.1–1,040	(ϕ 328)	250
Mercury	0.1– 10.2	(ϕ 2.9)	5

Sludges also contain carcinogenic substances such as 3,4 benzopyrene (400–1700 mg per kg compared with only 20 mg in normal agricultural soils) and may carry salmonella, worms and other parasites of man and animals. Careful and thorough fermentation in the treatment of sludge is needed to overcome this problem.

From the environmental point of view, the handling, transport and application of these materials causes a nuisance, particularly so in populated areas, though modern techniques such as aeration during fermentation can reduce this to tolerable levels.

Other organic fertilizers such as dried blood, bone meal and hoof and horn meal should be mentioned. These contain important plant nutrients, particularly nitrogen and phosphorus but only in relatively low concentration. They have a particular value in specialised horticulture but are not of significant importance in agriculture on account of high cost and limited availability.

References

Council for agricultural science and technology (1980) Organic and conventional farming compared. Report No. 84, pp. 32. CAST, AMes, Iowa.

Eich, D., Körschens, M. (1981) Zur Bedeutung der organischen Düngung für die Verbesserung der Bodeneigenschaften und zur Erzielung hoher und stabiler Erträge. Feldwirtschaft 22, 121–123.

Hinze, G. (1981) Möglichkeiten und Grenzen der Verwertung von Siedlungsabfällen für die Landwirtschaft unter besonderer Berücksichtigung von Müllkomposten und Müllklärschlammkomposten. Ber. Ldw. 59, 409–440.

Kampe, W. (1980) Bedeutung von Schwermetallen für den Pflanzenbau. Teil I: Situation, Bedeutung und Belastungsquellen; Teil II: Faktoren, Praxisaspekte und Rückläufigkeit der Belastung. Chem. und Techn. Landwirtsch. 31, 330–331 and 338–340.

Kloke, A. (1980) Die Bedeutung der organischen Düngung in der modernen Landwirtschaft. Gesunde Pflanzen 32, 185–187.

Oesterreichischer Wasserwirtschaftsverband (1980) Landwirtschaftliche Verwertung von Abwasserschlämmen. Dokumentation EAS-Seminar Basel, 24–26 September.

Paccolat, B. (1979) Agricultural use of sewage sludge. The Swiss experience. Phosphorus in Agriculture, No. 75, 29–38.

Pomel and Juste (Oct. 1977) Valorisation agricole des déchets T. 1 Composts urbains. INRA.

C 3 Alternative farming systems avoiding the use of fertilizers

Sections C 1 and C 2 have clearly demonstrated that supplies of plant nutrients from farm manures, sewage and other refuse and from legumes are insufficient to support full crop growth. There is certainly no possibility of increasing soil fertility by using these materials alone. Nevertheless, adherents of the bio-dynamic or organic schools never tire of proclaiming the virtues of alternative farming methods which prohibit or severely restrict the use of fertilizer even though trials of such methods show that yields are 10 to 40% below those on conventional farms using fertilizer. Using exclusively these organic materials does not confer any special properties on the soils treated with them (see Section B).

Plants obtain their nutrients from the soil. Native soil nutrients originate from the chemical and biological weathering of soil minerals and in the case of nitrogen from the breakdown of organic matter and from the atmosphere through fixation by free-living bacteria and by the root nodules of legumes. Up to 20 kg N per ha and small amounts of other nutrients may be added in rainfall. Small quantities are also added by dry deposition. Even the richest soils seldom supply sufficient of all nutrients for full crops.

There is no reason why "organically" treated crops should behave differently from those treated with fertilizer. Whatever may be the origin of plant food, plants can only take up nutrients in the ionic form, as they occur in natural soils. Nutrients in fertilizer occur as simple compounds which ionise when they dissolve in the soil solution. Nutrients in organic manures are present as complex compounds which have to be broken down into simpler form, a process which takes time. It can make no difference to the plant where the nutrient ions originate, but ions it must have.

The fact that time is required for nutrients in organic manures to become available can be advantageous in some circumstances because the slow release could give a steady supply of nutrient over a long period of growth. However, it is impossible to predict the rate of release, which is affected by weather and soil conditions, so it is impossible to control nutrient supply precisely. Fertilizer offers an advantage in that timing of application, splitting dressings etc. allows better control of the nutrient supply than can be attained with slow release materials. The possibility of placing fertilizer in the soil where it can do the most good enables good results to be obtained economically on the poorest soils.

Nobody would deny that farmyard manures, slurries and other waste materials are valuable and should be utilised in the interests of economical farming and conservation alike. But the quantities of these materials available are simply insufficient to allow economic optimum crop production and fertilizer must be used to make up the deficit. There follows a brief synopsis of some policies advocated by groups practising "alternative agriculture".

1. "Biodynamic farming"
Basis: "Landwirtschaftlicher Kursus" of the anthroposopher Rudolf Steiner, 1924 (Switzerland).
Preparations of horn meal, grated cow dung, "activated" with herbal additives and crushed quartz are supposed to promote the growth and quality of plants. Farmyard manure is composted with the help of herbal preparations. Within certain limits the use of commercial organic fertilizers is allowed (e.g. those based on bone meal, horn meal, dried blood, dried poultry droppings, etc., but not industrially produced nitrogenous fertilizers, readily soluble phosphate or potassium salts containing chloride. Rock phosphate, ground rocks and algal manures are acceptable.

Clover is the basic element of crop rotation. Weeds are removed mechanically. Special preparations and plant extracts are recommended for the control of fungal and animal pests. Special influences are attributed to cosmic forces whose beneficial effects are to be harnessed by arranging seed-time during specific constellations of the stars.

Products grown by biodynamic methods are marketed as quality products by the DEMETER Association under the label Demeter in FR Germany.

2. "Organic-biological farming"
Basis: Methods of Müller (Switzerland) and Rusch (FR Germany).
The chief emphasis of this form of agriculture is laid upon the promotion of micro-organisms in the soil and biocoenosis. This aim is served by minimum tillage and multiple cropping systems with preference for the leguminosae as interim crops. All water soluble mineral fertilizers are rejected, whereas basic slag and sulphate of potash magnesia are permitted. Nitrogen is only applied in organic form, primarily as farmyard manure, stable and liquid manure, compost, horn meal, dried blood and guano. The application of chemical pesticides is also forbidden. Only so-called "non-toxic" preparations can be used against fungal and animal pests. Weeds are controlled by mechanical and biological means i.e. by crop rotation.

Organic-biological food products are marketed in the FR Germany by the Bio-Land GmbH, under the trade mark "Bio-Land".

3. "Natural farming"
Basis: agricultural recommendations of the "Arbeitsgemeinschaft für naturgemässen Qualitätsanbau von Obst und Gemüse" (ANOG E.V.), according to Fürst.
Soil cultivation and fallow play an important role. The soil should be disturbed as little as possible but should be constantly covered with plants, including those for green manuring. Of the mineral fertilizers, phosphorus and potash fertilizers with sulphate content are permitted. Nitrogen-manuring is mainly organic, preference being given to dried blood, ricinus chips and similar manures.

24

For more demanding crops small amounts of mineral nitrates, such as Chilean saltpetre or calcium nitrate, are allowed.

Herbicides, fungicides and insecticides can be used selectively and to a limited extent, but the maximum values set by ANOG for the granting of certificates may not be exceeded.

It is calculated that at present the percentage of agricultural land farmed by all alternative systems is as shown in Table 3.1.

Table 3.1

Country	%
Belgium	0.03
FR Germany	0.10
England (without grassland)	0.04
France (depending on method considered)	0.03–0.24
The Netherlands	0.04

References

Aberg, E. (1976) Alternativa växtodlingsformer. Skogs- o. Lantbr.-akad. tidskr. 115, 315–331.
Böckenhoff, E. and Hertäg, O. (1981) Absatzbedingungen für alternativ wirtschaftende Betriebe. Ber. Landw. 59, 280–305.
Boeringa, R. (1980) Alternative methods of agriculture. Elsevier Sci. Publ., Amsterdam. pp. 200.
Boyeldieu, J. (1980) Organic farming and its prospects, compared with conventional farming. Phosphorus in Agriculture, No. 82, 31–38.
Brugger, G. and Held, R. (1980) Fragen der Wirtschaftlichkeit im alternativen Landbau. Kali-Briefe 15 (2), 109–121.
Dambroth, M., Wermke, M., Graff, O., Kloke, A., Diehl, J.F. (1978) Alternativen im Landbau – Statusbericht aus dem Forschungsbereich des BELF. Landwirtschaftsverlag Münster-Hiltrup, Reihe Landwirtschaft-Angewandte Wissenschaft, Heft 206, pp. 292.
Graf, U. and Keller, E.R. (1978) Zusammenhänge zwischen kosmischen Konstellationen und dem Ertrag landwirtschaftlicher Kulturpflanzen auf konventionell und biologisch-dynamisch bewirtschafteten Böden. Acker- u. Pflanzenbau 147, 40–59.
Mengel, K. (1979) Pflanzenbau ohne Mineraldüngung, eine Alternative? Kali-Briefe 14, 707–711.

C 4   Special properties of "organically" grown foods

Advocates of alternative farming contend that, in some pseudo-mystical way, organically grown food is "better" than that grown with fertilizer. There is no evidence that this is so (see Section J), neither is there any reason why it should be. Though there have been many long-term experiments which attempted to demonstrate this, it has never been possible to show that there is any validity in the arguments of the biodynamic school; neither has one hundred years of experience with fertilizer shown any sign of harmful effects.

As explained in section C 3 plants feed on ions so that the build-up of plant composition (carbohydrate, protein, fat, vitamins etc.) proceeds in exactly the same way whether nutrients are supplied in the organic or mineralised form. It is more possible to control plant composition to match a desired aim using fertilizer than by using organics because nutrient supply can be more precisely controlled (see Section C 3).

To quote examples of this, the juice of organically fertilized red beet had the extremely high nitrate content of approximately 2,900 mg per litre. Analysis of random samples of fresh spinach was unable to determine whether they originated from conventional or alternative farming systems. In a 15-year fertilizer trial the effects of mineral fertilizers and compost and stable manures on fodder and sugar beet, potatoes, lettuce, carrots, celery, spinach and savoy cabbage were compared; no significant differences were observed in the contents of the important nutritional components.

Other experiments have shown that the carotene and vitamin (B1, B2, C) contents of crops were not reduced by the use of mineral fertilizers; in fact mineral fertilizers used in addition to farmyard manures have increased the carotene level as well as calcium, phosphorus, iron and copper contents in carrots and spinach. The baking quality of wheat receiving mineral fertilizers is at least as good as that obtained with organic manures, and can in fact be improved by the judicious use of N fertilizer. The question of taste is open to subjective judgement but in one experiment comparing the taste of mineral-fertilized and biodynamically fertilized potatoes, the former was considered to be superior. So far, there is absolutely no indication that food grown without mineral fertilization is of higher nutritional value than other food. The opinion held by some people that food products obtained by "alternative" methods are of better quality than other food produced by using mineral fertilizers is not therefore supported by fact.

ment of Nutritional Sciences, University of California "the substitution of 'organic' and 'chemical' fertilizers during the growth of plants produces no change in the nutritional or chemical properties of foods. All foods are made of 'chemicals'".

26

References

Anonymous (1979) Düngen wir richtig? Kurzbericht über VDLUFA-Kongress Augsburg 1978, Verlagsgesellschaft Ackerbau, Kassel, 1–32.

Diehl, J.F. and Wedler, A. (1978) Vergleichende Untersuchungen über die Qualität der Ernteprodukte. Landwirtsch.Angw. Wissenschaft, Heft 206, Alternativen im Landbau, Landwirtsch.-Verlag, Münster-Hiltrup, 150–292.

Jukes, T.H. (1977) Organic Food. Critical Reviews in Food Science and Nutrition 9, 395–418. CRC Press, Boca Raton, Florida.

Schuphan, W. (1974) Ertrag und Nahrungsqualität pflanzlicher Erzeugnisse unter besonderer Berücksichtigung der Problematik "organischer" oder "chemisch-mineralische" Düngung. Ernährungs-Umschau 21, 103–108.

Vetter, H. (1980) Umwelt und Nahrungsqualität. Heyne-Verlag München, 1–110.

D Crop quality

D 1 Definition of crop quality

Quality comprises all those properties of the harvested product that contribute to its value to the consumer. These include:
1. external qualities − colour, appearance and grading,
2. compositional qualities all dependent on chemical composition − taste and flavour, energy-value (i.e. calorie content), protein content, biological value of the protein, mineral content, vitamin content,
3. storage and transportation qualities (for storage of varying duration before processing and/or marketing), and
4. processing and cooking qualities.
The various properties are not necessarily interrelated. Thus a product which is regarded as having good compositional qualities will not necessarily be of a high quality for processing, and so on. Equally, external factors such as soil, climate and methods of cultivation and postharvest handling will not necessarily affect the different aspects of quality in the same way.

References

Howard, H.W. and Hughes, J.C. (1974) Factors influencing the quality of ware potatoes. 1) The genotype. 2) Environmental factors. Potato Res. 17, 1) 490−511, 2) 512−547.
Jahn-Deesbach, W. (1975) Zum Qualitätsbegriff bei pflanzlichen Rohstoffen. Ergebnisse landw. Forschung a.d. Justus-Liebig-Universität Giessen, Heft 13, 9−17.
Kampe, W. (1981) Agrarchemikalien und Nahrungsqualität. Der Stickstoff No. 13, 14−32.
Leitzmann, C., Fuss, E. and Meier-Ploeger, A. (1979) Ernährungsphysiologische Qualität von Obst und Gemüse. Ernährungs-Umschau 26, 286−293.
Kübler, W. (1976) Die Bedeutung pflanzlicher Nahrungsmittel für die Deckung des Bedarfs an Vitaminen und Mineralstoffen. Ernährungs-Umschau 23, 107−110.
Miflin, B.J. (1975) Potential for improvement of quantity and quality of plant proteins through scientific research. Proc. 11. Colloq. Internat. Potash Inst., Rønne, Denmark, 53−74.
Nilsson, T. (Ed.) (1979) Symposium on quality of vegetables. Acta Horticulturae, No. 93, pp. 444.
Röbbelen, G. (1976) Besseres Pflanzeneiweiss durch Anbaumassnahmen und Züchtung. Ernährungs-Umschau 23, 49−57.
Solms, J. (1977) Sensory qualities of carbohydrates and lipids. Proc. 13. Colloq. Internat. Potash Inst., York, U.K., 157−167.
Trenkle, K. (1976) Qualität, Lebensmittelqualität, Nahrungsqualität. Verbraucherdienst 21, 177−180.

D 2 Effect of nitrogenous, phosphatic and potassic fertilizers on quality

Interaction with other factors often makes it difficult to obtain a clear insight into fertilizer effects on quality factors of plant products.

There is no doubt that high quality products can only be obtained if N, P and K (and other essential elements) are supplied in sufficient and correctly balanced proportions. The influence exerted by the individual elements on growth, composition and quality of plant products is, however, very complex.

Each of the essential nutrients is involved – at least as a component in maintaining the correct balance – in the growth mechanism and physiological processes which ultimately determine crop yield *and* quality. The individual nutrients have specific beneficial effects, either singly or by mutual inter-action, on the synthesis of plant components and thus on the various aspects of crop quality.

The major effects of the three primary fertilizer nutrients on crop quality are briefly summarized below (for further details see Sections D, E, F and J). The effects are the same whether the nutrients come from fertilizers or from organic manures.

Nitrogen (N) favours:

Vigour of plant growth. It accelerates the development of above- and below-ground parts of the plant and influences the size, weight and colour of crop products, e.g. grains, tubers and fruits.

Synthesis of amino acids, proteins and lipids and so, to a large extent, the nutritional value of food products.

Technological qualities related to protein content, e.g. baking quality of cereals.

Phosphorus (P) favours:

Synthesis of organic phosphorus compounds in plants, e.g. phospholipids, nucleic acids and enzymes, which are essential to the metabolic formation of proteins, carbohydrates and other lipids and enzymes.

The reproductive phase of the plant, i.e. maturity and quality of fruit and seed.

Mineral status of food and fodder crops.

Potassium (K) favours:

Almost all metabolic processes involved in the synthesis of valuable constituents such as proteins and carbohydrates, because it functions as an activator of enzymes essential to metabolism and interacts with P and N.

Vitamin and mineral contents, especially in vegetables and fruit.

Technological qualities dependent on carbohydrate content, e.g. processing quality of potatoes, malting quality of barley.

Both P and K favour:

Photosynthesis and translocation of photosynthates from leaves to storage tissues.
Resistance to pests and diseases.
Hardiness against cold and drought.
Resistance to lodging.
Handling and storage properties of harvested produce.
Provision of an unbalanced supply of one or more of the major nutrients is liable to diminish these beneficial effects and may lead to an upset in plant metabolism. Thus, it is only when fertilizers are incorrectly used that the quality of plant produce is impaired.

References

Amberger, A. (1978) Mineralische Düngung und Qualität der Nahrungsmittel. Boden-kultur 29, 132–139.
Buchner, A. and Sturm, H. (1980) Gezielter düngen. DLG-Verl. Frankfurt/M., pp. 319.
Chemical Society, U.K. (1979) Chemistry and agriculture. Spec. Publ. No. 36, pp. 287.
FAO/ECE (1974) Symposium on the effects of fertilizers on the quality and nutritional value of grains, potatoes, selected fruits and vegetables, and forages. Geneva/Switzerland.
Glatzel, H. (1978) Mineraldüngung und Pflanzenqualität. Der Stickstoff No. 12, 5–13.
Greenwood, D.J., Cleaver, T.J., Turner, M.K., Hunt, J., Niendorf, K.B. and Loquens, S.M.H. (1980) Comparison on the effect of K-, P- and N-fertilizer on the yield, K-, P- and N-content and quality of 22 different vegetable and agricultural crops. J. agric. Sci. Camb. 95, 441–485.
Handbook on Phosphate Fertilization (1981) ISMA, Paris.
Kuntze, H. and Voss, W. (1980) Statusbericht Düngung. Landwirtschaftsverl. Münster-Hiltrup, pp. 144.
Munson, R.D. (1976) Role of fertilizer nutrients in crop quality. Commun. Soil Sci. Pl. Analysis 7, 497–511.
VDLUFA (1979) Düngen wir richtig im Blick auf Pflanzenqualität? 90. Kongr. Augsburg 1978, Landw. Forschung, So.heft 35, 1–132.
Vetter, H. (1980) Umwelt und Nahrungsqualität. W. Heyne Verlag, München, pp. 110.
Zehler, E., Kreipe, H. and Gething, P.A. (1981) Potassium sulphate and potassium chloride. Their influence on the yield and quality of cultivated plants. Internat. Potash Inst., Research Topics No. 9, pp. 108.

D 3 Cereals

Many characteristics of cereal quality can be influenced by nutrient supply.

The concept of quality depends on the requirements and preferences of users, on the purpose for which the product is used, and on nutritional value. Aspects to be considered in the case of cereals may be divided into:
1. external characteristics (1,000-grain weight, hectolitre-weight, etc.),
2. internal characteristics (content of proteins, essential amino acids, unsaturated fatty acids, carbohydrates, vitamins and minerals), and
3. technological properties (suitability for milling, baking, biscuit-making, pasta manufacture or malting).

The content and yield of protein are of special importance. Cereals at the present time account for over 50% of human consumption of protein in the world as a whole. Since protein is a limiting factor in both human and animal nutrition, and the world population is growing rapidly, every effort should be made to increase the yield and quality of cereal protein.

The primary plant nutrients nitrogen (N), phosphorus (P) and potassium (K) influence cereal quality in different ways.

For a summary of effects on individual cereal crops the reader is referred to the FAO paper by Ufer (1974) which includes a bibliography of over 700 references. The following remarks are, of necessity, generalizations.

External characteristics

The effects of nitrogen depend on climate and other factors. The timing of fertilizer nitrogen application in relation to the stage of growth of the crop and the incidence of rainfall after fertilizer application are of particular significance. Thus, depending on weather conditions in the season in question and on the rate and timing of fertilizer application, the 1,000-grain weight may be either increased or decreased, or unaffected, by nitrogen. In general, early application tends, if anything, to reduce the 1,000-grain weight and late application to increase it. These effects are probably more marked under the more consistent weather conditions of a continental climate than under the the very variable conditions of the maritime climates of, for example, the British Isles and Norway. Thus, in a number of countries, split application of nitrogen is recommended, including a dressing during the heading and flowering stages, not only to improve the internal characteristics of the grain and to reduce the risk of lodging, but also to encourage grain formation and to maintain or increase the 1,000-grain weight.

Phosphorus increases the proportion of large grains, and improves the ratio of grain to straw. Large numbers of field trials on responsive soils have demonstrated favourable effects on 1,000-grain-weight (wheat, rye, barley, oats) and on hectolitre weight (barley, maize).

Potassium, like phosphorus, increases the proportion of large grains and has favourable effects, on responsive soils, on 1,000-grain and hectolitre weights.

Internal characteristics

Nitrogen topdressing, particularly when applied during the heading and flowering stages, increases both percentage of protein in the grain and the total yield of protein. Indications are that the relative proportions of the essential amino acids leucine and phenylaniline increase while those of lysine, threonine, trytophane, cystine, arginine, aspartic acid and glycine decrease. It should be noted, however, that a smaller proportion of an individual amino acid such as lysine, in the total protein, does not necessarily imply a lower quantity of this amino acid in the grain; it may indeed be larger.

Phosphorus is of particular importance for the vitality of the seed, as evidenced by the better germination capacity of grain from cereals well supplied with this nutrient. Potassium, when balanced in the correct proportions by phosphorus and nitrogen, extends the period of grain filling and maturation. In addition to increasing the potassium content of the grain, potassium also increases the crude protein content and affects the relative proportions of the different protein fractions (albumin, globulin, gliadin, glutenin, etc.) and individual amino acids. Deficiency of potassium, for example, increases glutamic acid and proline relative to other amino acids. In particular, an adequate supply of potassium can increase the proportion of lysine in the grain.

Technological properties

Wheat. The baking quality of wheat for breadmaking depends on gluten quality. The quantity of gluten is generally increased by late nitrogen topdressing, particularly at the heading and flowering stages. This is of special importance in the case of hard wheats for pasta manufacture. Phosphorus and potassium alone reduce the gluten content, but this is compensated by nitrogen, so that a complete NPK treatment will, in general, result in an increase in gluten. For the best baking quality it is necessary to have a high content of gluten of good quality. Provided the wheat variety grown is of good quality, in that it contains good quality gluten, its baking quality will be considerably improved by appropriate fertilizer treatment, including late nitrogen topdressing. The baking quality of poorer quality wheats, on the other hand, which contain gluten of inferior quality, will not be improved to the same extent. Under some conditions the increased protein content resulting from fertilizer treatment, which is associated with improved baking quality, may also be associated with factors such as colour and alpha-amylase activity which are regarded as undesirable by millers producing white flour for bread manufacture on a large scale. Further research is needed on this

32

subject. The precise requirements for good quality will obviously depend on the scale of operations, on economic factors and on the types of flour and bread that local consumers demand.

Rye. Unlike wheat, the milling and baking quality of rye for breadmaking is associated with the carbohydrate constituents, particularly the pentosans. In general, there appears to be no relationship between fertilizer use and baking quality.

Barley. Malting quality of barley is adversely affected by excessive or late application of nitrogen which brings the protein content of the grain above the desired level, and it is improved by phosphorus and potassium. Adequate potassium has a mostly favourable effect on malt solubility, clarification time, colour and residual extract of beer. For the particular case of whisky-making (other than pure malt whiskies), a high diastatic power in the barley malt is considered essential; there is evidence that this is increased by fertilizer nitrogen application.

Feed grains. The nutritional quality of feed grains and of cereals grazed or cut for silage is improved by appropriate fertilizer treatment, through increase in energy value, and in protein content and quality, and in mineral content.

References

Beringer, H. and Haeder, H.E. (1981) Influence of potassium nutrition on starch synthesis in barley grains. Z. Pflanzenernähr. Bodenkd., 144, 1–7.

Bolling, H. and Seibel, W. (1978) Einfluss der Düngung auf die Getreide-qualität und den Nährwert der Lebensmittel auf Getreidebasis. Der Stickstoff No. 12, 23–30.

Buchner, A. and Sturm, H. (1980) Gezielter Düngen. DLG-Verlag Frankfurt/M., pp. 319.

Fajersson, F. (1974) Climate, fertilisation, variety – essential factors for wheat quality. Phosphorus in Agriculture, No. 62, 49–59.

International Potash Institute (1975) Fertilizer use and protein production. Proc. 11. Colloq., Rønne, Denmark, 243–306.

Stewart, B.A. (1977) The effects of fertilizers and other agricultural inputs on quality criteria of wheat needed for milling and baking. Proc. 13. Colloq., Internat. Potash Inst., York, U.K., 243–250.

Ufer, M. (1974) The influence of fertilizers on the quality of cereals. FAO/ECE Symp. on the effects of fertilizers on the quality and nutritional value, Geneva, June.

Wirths, W. (1974) Optimale Düngung verbessert ernährungsphysiologische Qualität. Umschau Wiss. Techn. 74, 208–212.

Zoschke, M., Fischbeck, G., Bolling, H., Seibel, W. and Teuteberg, W. (1979) Düngen wir richtig im Blick auf die Qualität der Körnerfrüchte. 90. VDLUFA Kongr. Augsburg 1978, Landw. Forschung, So.heft 35, 43–53.

D 4   Root crops

Sugar Beet: Technological quality of sugar beet is determined both by the conditions under which the crop is grown and by the effects that fertilizer treatment may have on the physiology of the crop. As with cereals, the quality characteristics to be considered may be divided into:
1. external (root shape, etc.),
2. internal (sugar content, juice purity, soluble ash, etc.).

External characteristics. Nitrogen within the recommended economic range of usage, together with adequate P and K, encourages root growth but has an even greater effect on top growth. Phosphorus is particularly important for crops grown for seed, improving seed size and quality. Fertilizer treatment has little effect on root shape or ease of slicing.

Internal characteristics. The following general observations are taken from the FAO report by Ufer (1974):

"*Sucrose content* was generally accepted as least variable in relation to fertilizers ... N fertilizers at too high level sometimes decrease the sucrose content, but in other tests sucrose is the only constituent which increases with increased rates of N fertlizers ... Under normal conditions there is generally no significant influence of P and K on sucrose percentage, but, if there is, K seems of slightly more effect in increasing sugar content than P. There are very contradictory statements on the influence of micro-elements ... It seems that under favourable conditions some trace elements added to basic NPK fertilizers will increase the content of sucrose. On *juice purity*, P and K generally have little effect. Excessive potassium and also sodium (applied as nitrate or in other form) reduce juice quality. Some trace elements, such as boron, copper and manganese, affect juice purity favourably.

Noxious nitrogen, which decreases the sugar output by increasing the proportion of molasses, is in direct relation to maturity ... A good balanced NPK fertilization provides mature roots and prevents the formation of too much noxious nitrogen. If N is too high in relation to P and K, noxious nitrogen increases ... K consistently diminishes noxious nitrogen (by reducing the concentrations of α-amino N and molasses-forming N and by favouring rapid translocation of sugar to the roots) ... The influence of trace elements is uncertain ..."

There is evidence that the sucrose content is higher with denser planting and when the crop is irrigated. Irrigation also reduces noxious nitrogen. Official recommendations for fertilizer treatment to give optimum yield of acceptable quality vary appreciably from country to country. This is no doubt due in part to differences in soils and climate conditions, but also to differences in traditional attitudes and practices.

Because of these differences, it is not practicable to indicate generally applicable critical levels of fertilizer nitrogen in respect of sucrose content and noxious nitrogen level. As an example it may be stated that an excess of 75 kg N per ha above the recommended level may result in both the sugar percentage and the juice purity being about 0.5 lower than if the recommended amount of nitrogen had been applied. (In the example on which these figures are based, the sugar content and juice purity corresponding to the recommended N dressing were about 16.6% and 92.3% respectively).

Potatoes: The criteria of quality depend on end-use.
Thus, one may differentiate:
1. seed potatoes,
2. potatoes for human consumption,
   (a) directly (early or main crop)
   (b) after processing (e.g. potato crisps, frozen, dehydrated and canned potatoes,
3. feed potatoes for consumption by livestock, and
4. industrial potatoes for manufacture of starch, alcohol and other products.

As with sugar beet, the quality characteristics to be considered may be divided into:
1. external (tuber size distribution, shape, specific gravity, external appearance, resistance to mechanical damage),
2. internal (texture, colour, flavour and contents of starch, sugars, protein, individual amino-acids, vitamin C, etc.).

External characteristics: In general, nitrogen and potassium increase tuber size and the proportion of large tubers, while phosphorus increases the yield of medium-sized rather than large tubers and thus exerts a favourable influence on grading.

Potassium deficiency results in longish tubers small in diameter. Potassium sulphate gives smaller tubers than potassium chloride, an advantage for certain uses (e.g. seed potatoes, and processing for human consumption). Specific gravity, a factor which is taken into account in the price of industrial potatoes, is adversely affected by excessive nitrogen, i.e. in excess of the level which gives a reasonable response in tuber yield. Phosphorus increases specific gravity where there is a yield response to phosphorus. Phosphorus also promotes development and maturity, and improves the sprouting capacity of seed potatoes. By strengthening the skin, both phosphorus and potassium increase resistance to mechanical damage during harvesting, transportation and storage.

Internal characteristics: Potatoes for direct human consumption

Overall, there is little concern on the effects of fertilizers on potato quality (which are smaller than the effects of site and season) provided that appropriate nutrient rates are used:

"Good balanced N, P and K fertilization generally increases dry weight, but lowers starch content and sugars of tubers. Protein content and vitamin C content rise significantly. If N fertilizing is too high in relation to P and K, crude protein increases but the tendency of the potato to darken during cooking is intensified. Topdressing of N seems to increase especially the starch percentage of dry matter, but to a small degree also proteins and vitamin C. Single application of N has better results on table quality than split application ... Texture, flavour and taste determine the quality of the table potato. Colour is less important. Increasing N fertilization seems to increase the tendency to discoloration (from bruising), while K counteracts it ... Post-cooking blackening, which is mostly due to inadequate K fertilization ... is associated with the production of melanine through enzymatic oxidation of tyrosine. Free tyrosine in the tuber is increased by raising N fertilization and short supply of K. Therefore, an adequate rate of K should prevent the increase of tyrosine by N. Mg intensifies the effects of K, while P has no influence ... The level and source of K affects the texture ... the optimum level and kind depends mostly on the soil. K sulphate tends to raise the dry weight of the tuber and makes it more floury ... K sulphates, in particular, are important for good cooking and chipping quality ... There are other experiments which show only small influences of fertilizers compared with those of climate, season and soil ... Chloride fertilizers are said to lower the vitamin C content, whereas sulphates mostly cause it to increase ... These examples demonstrate the complexity of the fertilizer-quality relationship of table potatoes. It is considered to be one of the most difficult problems of quality research." (Ufer 1974)

Potatoes for processing for human consumption. The industry prefers potatoes with a high dry matter content. For this reason, potassium sulphate tends to be favoured in preference to potassium chloride. Another desired quality is a low content of reducing sugars; this applies in particular to chips and crisps. The reducing sugar content can be affected by many factors, including fertilizers, but it is difficult to draw general conclusions. Nitrogen, for example, has been reported to affect the reducing sugar content adversely if the haulms are still vigorous at the time of harvesting, yet under other conditions it has had a beneficial effect. Similarly, potassium is considered by some authors to result in a lower content of reducing sugars, but there is a divergence of opinion on this point. Potassium is of major importance for reducing discoloration (see above).

Feed potatoes. Quality is usually a secondary consideration since it is mainly surplus or unsaleable potatoes that go for animal feed.

Industrial potatoes. The starch industry requires potatoes with a high starch content of good quality (amylose) and with viscosity. Phosphorus improves starch quality and viscosity, while potassium increases the proportion of larger starch granules; this is a desirable characteristic, for the larger granules gelatinize more easily. The proportions of amylose and amylopectin are unimpaired by normal rates of N, P and K, but an excess of N or a deficiency of K may significantly lower the amylose content.

References

Burba, M.T. (1980) Kalium und Natrium im Stoffwechsel der Zuckerrübe. Zuckerrübe 29, 16–19.

Carter, J.N., Westermann, D.T. and Jensen, M.E. (1976) Sugarbeet yield and quality as affected by N Level. Agron. J. 68, 49–55.

Conesa, A.P., Mettauer, H., Vuittenez, A. and Trenkel, R. (1980) The effect of agronomic factors on processing quality of sugar beet. Phosphorus Agric. 34, No. 79, 3–15.

Draycott, A.P., Durrant, M.J. and Last, P.J. (1974) Effect of fertilizers on sugar quality. Internat. Sugar J. 76, 355–358.

Holsbeek, W.F. and Scheys, J. (1981) L'aptitude du sol et les conditions nutritives optimales pour obtenir des betteraves sucrières de qualité. Revue de l'Agriculture 34, 17–33.

Hunnius, W., Müller, K. and Winner, C. (1979) Düngen wir richtig im Blick auf die Qualität von Hackfrüchten? 90. VDLUFA-Kongr. Augsburg 1978. Landw. Forschung, So.heft 35, 54–71.

International Potash Institute (1977) Effects of fertilizers on the production of carbohydrates. Proc. 13. Colloq., York, U.K., 175–227.

Ludwick, A.E., Gilbert, W.A. and Westfall, D.G. (1980) Sugarbeet quality as related to KCl fertilization. Agron. J. 72, 453–456.

Marschner, H. and Krauss, A. (1980) Beziehungen zwischen Kaliumgehalt und Qualität von Kartoffeln. Kartoffelbau 31, 65–67.

Ufer, M. (1974) Effects of fertilizers on the quality of potatoes and root crops. FAO/ ECE Symp. on the effects of fertilizers on the quality and nutritional value, Geneva, June.

Wegener, J., Ziegler, G. and Sternkopf, G. (1979) 1. Untersuchungen zur Schwarzfleckigkeit von Kartoffelknollen in Abhängigkeit von N-, Stallmist- und K-Düngung sowie Zusatzberegnung und Reifegrad der Knollen. 2. Uber Zusammenhänge der Schwarzfleckigkeit von Kartoffelknollen mit dem Trockenmassegehalt und den Inhaltskonzentrationen sowie mit der Düngung. Arch. Acker- u. Pflanzenb. Bodenkd. 23, 297–314.

D 5   Oilseed crops

Successful production of crops which yield oil of high nutritional value and of good quality for industrial processing depends — to an even greater extent than that of other crops — on an abundant but correctly balanced supply of all plant nutrients, especially K and N.

Crops such as oilseed rape, linseed, sunflower, olives, nuts and soya beans are producers of proteins, vitamins and essential fatty acids and are rich in lipids. When considering the effect of fertilizers on the oil content of these crops, the qualitative composition of saturated and unsaturated fatty acids must also be taken into account. Saturated fatty acids (such as palmitic and stearic acids) have single C-bondings and give firm fats of high technical value. Unsaturated fatty acids (the most essential of which are oleic, linoleic, linolenic and arachidonic acids) can only be synthesized by plants; their high physiological value as energy sources for man and animals results from their reactive double C-bondings.

Maximum oil production depends on conditions which favour photosynthesis. The resultant photosynthates can be utilized *via* two alternative pathways, one leading to the production of proteins and the other to the synthesis of fats. If ample NH_4 is available, protein production is favoured and less carbohydrates are available for synthesis of fats. If, on the other hand, there is a poor supply of NH_4, then synthesis of fats is favoured. Many experiments with oilseed crops (e.g. sunflower, rape) and cereals (especially oats) have demonstrated the importance of a sufficient and balanced supply of nitrogen and potassium for fat synthesis in the plant.

Restricted vegetative growth of plants is usually characterized by high carbohydrate and oil contents. Under conditions of high respiration activity and increased vegetative growth, which can result in particular from heavy use of nitrogen unbalanced by phosphorus and potassium, less carbohydrate is available for fat synthesis; so the oil content decreases, and there are more empty seeds. The increased seed yield resulting from nitrogen use in general, however, outweighs the lower oil content, so producing higher oil yield. Potassium can assist by maintaining a favourable ratio in the plants between carbohydrate and nitrogen.

The beneficial effect of soil potassium and of potassium fertilizers on the oil content of fruits and seeds is generally acknowledged. In olive cultivation, for example, abundant NK supply leads to the production of a greater number of fruits of an acceptable size for canning and the increased yields of oil with a high content of essential unsaturated fatty acids (e.g. oleic and linoleic acids) and a low content of saturated acids (e.g. palmitic acid). Increased potassium in the fruits is correlated with a higher oil content. Fermentation is improved by balanced NK fertilization. In Italy, the best olive fruits for fresh consumption are grown on soils which are rich in both phosphorus and potassium. The seed yield and oil content of soya beans and sunflowers are

38

similarly increased by a balanced supply of plant nutrients, particularly phosphorus and potassium.

Whilst boron plays a key role in the process of oilseed formation, an ample supply of sulphur not only affects the yield but also increases the content of sulphur-containing glucosides, components which, e.g. in rapeseed, are not wanted for processing. The content of undesirable constituents such as erucic acid, in rape seed oil of the older varieties, is unaffected by fertilizer treatment.

References

Coic, Y. (1975) Cumulative and antagonistic effects of fertilizers on the protein and oil contents of dual purpose pulses. Proc. 11. Colloq. Intern. Potash Inst., Rø nne, Denmark, 161–168.

Holmes, M.R.J. (1980) Nutrition of the oilseed rape crop. Applied Sci. Publ., London, pp. 158.

International Potash Institute (1977) Effects of fertilizers on the production of lipids. Proc. 13. Colloq. York, U.K., 257–340.

Kumar, V., Singh, M. and Singh, N. (1981) Effect of sulphate, phosphate and molybdate application on quality of soybean grain. Plant and Soil 59, 3–8.

Samui, R.C. and Bhattacharyya, P. (1980) Effect of soil and foliar application of N, K and Mo on oil content and yield and chemical composition of sunflower. J. Indian Soc. Soil Sci. 28, 193–198.

Scott, R.K., Ogunremi, E.A., Ivins, J.D. and Mendham, N.J. (1973) The effect of fertilizers and harvest date on growth and yield of oilseed rape sown in autumn and spring. J. agric. Sci. Camb. 81, 287–293.

Teuteberg, W. and Trautschold, E.W. (1979) Effect of harvest time and fertilization on the components of rape. Proc. 5. Internat. Rapeseed Conf. 1978, 235–244.

D 6   Fruit and vegetables

As a part of man's higher standard of living, fruit and vegetables play an increasingly important part in his daily diet. Appearance, flavour and texture, and suitability for canning and processing, are key qualities in marketability. Nutritional value depends mainly on vitamin and mineral contents. They are particularly important sources of potassium and magnesium. Some types play an important role in maintaining a correct balance between sodium and potassium in the diet.

These various qualities are beneficially affected by fertilizers in many ways, both through the separate effects of the individual nutrients and by interactions between them. In general, the quality of fruit and vegetables depends very much on the correct balance and not only on the total amount of nutrients applied.

Although only the effects of the major plant nutrients nitrogen, phosphorus and potassium are considered below, the secondary and micro-nutrients are also important in relation to quality: for example, magnesium and boron for fruit crops and sulphur for cruciferous vegetables.

Characteristics affecting marketability

Fruit

Adequate fertilizer, depending on species, soil and intensity of cropping, (in high-yielding commercial orchards, around 140–200 kg N, 75–100 kg P_2O_5, 200–300 kg K_2O per ha per year) applied in correctly-timed split dressings, improves fruit set and reduces "June drop"; coupled with effective pruning, fruit thinning and soil maintenance, it ensures a good crop and a high individual fruit weight. Deficiency of nitrogen, seldom encountered in commercial orchards, not only results in small-sized fruit but the colour may change too soon and the desired fresh, succulent taste may be lacking. Excessive nitrogen — i.e. an unbalanced N : P ratio — can reduce the firmness of the flesh of apples, pears and peaches and, in the case of citrus fruits, result in a thick, coarse peel, more normally associated with variable weather. A balanced N : P ratio favourably affects flavour (sugar : acid ratio), e.g. of apples and wine grapes. Potassium fertilizer is especially important in relation to maturity, fruit colour, flavour and storage quality.

Vegetables

By promoting rapid and succulent leaf growth, N-fertilization has obvious advantages for leaf vegetables such as lettuce and cabbage. Phosphorus fertilization not only improves head formation of lettuce and cabbage but also firmness of the head and its keeping quality in winter or cold storage. Phosphorus is also reported to have improved the flavour rating of canned carrots, peas and asparagus, and to have increased the size and quality grading of onions, celery, lettuce and spinach. Potassium fertilization improves the firmness and resistance to damage during transport and storage, of many vegetables and promotes thick growth of carrots and improved size and shape of beetroot. The ripening of tomatoes is speeded by application of phosphorus. Potassium aids uniformity in shape, ripening and colouring of the fruit; high levels of this nutrient in soil and plant prevent ripening disorders such as "greenback" and "hollow fruit".

Nutritional characteristics

N fertilization enhances the synthesis of carbohydrates and proteins in vegetables such as spinach, cabbage, celery, etc. Balanced NPK application increases the content of total soluble solids, an important criterion for citrus fruit or tomatoes for processing. Sugars, which contribute with fruit acids to taste and flavour, are increased by phosphorus and potassium, e.g. in apples, peaches, carrots, onions and spinach. Potassium lowers the proportion of reducing sugars relative to sucrose. An intimate relationship between potassium and/or nitrogen and organic acids is found in fruits and tomatoes.

As regards vitamins, nitrogen and phosphorus markedly increase the contents of carotene (precursor of vitamin A) and vitamins B1 and B2, particularly in leafy vegetables, carrots and tomatoes. Potassium increases most carotenoids in tomatoes, and especially ascorbic acid (vitamin C), which is also increased in apples by nitrogen. (See also Section J 3).

Fertilizers promote the mineral content of both fruit and vegetables – vital for human metabolism.

References

Adams, P., Davies, J.N. and Winsor, G.W. (1978) Effects of N, K and Mg on the quality and chemical composition of tomatoes grown in peat. J. hort. Sci. 53, 115–122.

Atkinson, D., Jackson, J.E., Sharples, R.O. and Waller, W.M. (Eds.) (1980) Mineral nutrition of fruit trees. Butterworths London-Boston, pp. 435.

Bünemann, G. (1981) Nährstoffversorgung und Qualitätserzeugung bei Obst. Der Stickstoff No. 12, 39–51.

Fritz, D. (1980) Probleme bei der Erzeugung von Gemüse im Gartenbau und Landwirtschaft. Ernährungs-Umschau 27, 182–188.

Fritz, D. (1981) Einfluss der Mineraldüngung auf die Qualität von Gemüse. Der Stickstoff No. 12, 14–22.

Gautier, Ph., Gagnard, J. and Thiault, J. (1975) Etude de l'influence des caractéristiques des vergers et la nutrition des arbres sur la qualité gustative des pommes. Bull. techn. Inf., Paris. No. 301, 425–440.

Greenwood, D.J., Barnes, A., Kit Liu, Hunt, J., Cleaver, T.J. and Loquens, S.M.H. (1980) Relationships between the critical concentrations of N, P and K in different vegetable crops and duration of growth. J. Sci. Food Agric. 31, 1343–1353.

Jungk, A., Paulus, K. and Seibel, W. et al. (1976) 1. Beeinflussung des Vitamin- und Mineralstoffgehaltes von Pflanzen durch Züchtung und Anbaumassnahmen. 2. Veränderungen des Vitamin- und Mineralstoffgehaltes von Nahrungspflanzen durch technologische Massnahmen. Ernährungs-Umschau 23, 111–115, 116–123.

Mengel, K. (1979) Influence of exogenous factors on the quality and chemical composition of vegetables. Acta Horticulturae No. 93, 133–151.

D 7   Pests and diseases

Susceptibility to disease or pest attack depends both on the plant variety and on the environment in which it is grown. Plants receiving adequate and correctly balanced fertilizer treatment are in general less susceptible to diseases and pests than plants with an inadequate or unbalanced nutrient supply. Potatoes can better overcome the effects of phythophthora when adequately supplied with nitrogen. By permitting more intensive cropping, or by influencing the height and size of individual plants, fertilizer use may, however, have an indirect effect on the spread of pests and diseases.

The effects of the different nutrients vary widely, because the incidence of some organisms is more severe on slow-growing, weak hosts, whilst that of other organisms is more severe on thriving plants. Nitrogen, for example, can enable young plants to grow away from soil-borne fungal diseases, e.g. root and stem rots. Phosphorus similarly reduces the incidence of root diseases by promoting rapid growth of roots and facilitating quick recovery of affected roots; and potassium, by strengthening cell walls and promoting firmer tissue and stronger stems, reduces incidence of many stem and leaf diseases.

High potassium and nitrogen supply promotes the synthesis of high molecular compounds so that the pathogens are deprived of soluble N compounds and low molecular carbohydrates as nutritional source.

It is noteworthy that after 120 years of almost continuous spring barley in one of the long-term experiments at Rothamsted the fully fertilized plots receiving N, P and K are still yielding well and suffer little damage from root rots.

References

Beringer, H. and Koch, K. (1980) 1. Einfluss von K-Ernährung und Mehltaubefall auf den N-Haushalt von Sommergerste während des Kornwachstums. 2. Aufnahme und Verlagerung von 15N-Stickstoff bei Sommergerste in Abhängigkeit von K-Ernährung und Mehltaubefall. Z. Pflanzenernähr. Bodenkd. 143, 441–456.

Bockmann, H. and Partsch, G. (1975) Zusammenhänge zwischen N-Düngung und Pflanzenkrankheiten im intensiven Getreidebau. BASF-Mitt. für den Landbau, No. 2, pp. 58.

Böning, K. (1973) Uber die Beziehungen zwischen Ernährung der Pflanze und ihrer Anfälligkeit für parasitäre und nicht parasitäre Krankheiten. Mitt. BBH Land- Forstwirtsch. Berlin, Heft 151, pp. 15.

Dimitri, L. (1978) Einfluss der Düngung auf die Gesundheit der Waldbestände. Allg. Forstz. 15, 411–413.

Internat. Potato Center (1976) Fertilizer use and plant health. Proc. 12. Colloq., Izmir, Turkey, pp. 330.

Internat. Potato Center (1978) Control of important nematode and insect pests of potatoes. Ann. Report, Lima, Peru, 33–41.

Kolbe, W. (1977) Untersuchungen über den Einfluss mineralischer und organischer Düngung auf den Krankheitsbefall im Obstbau bei einheitlichen Pflanzenschutzmassnahmen (1961–76) Pfl.schutz-Nachr. BAYER 30, 138–152.

Kurkela, T. (1975) Evaluation of the effects of forest fertilization on damages caused by diseases. 2. World Techn. Consult. Forest Diseases and Insects. New Delhi, FAO-Dok. No. 29263.

Perrenoud, S. (1977) Potassium and plant health. Internat. Potash Inst., Research Topics No. 3, pp. 218.

44

D 8   Frost and drought

In general, fertilizers improve the resistance of plants to drought and also to frost, since this is related to the water status of the whole plant and to the turgidity and permeability of the plant cells.

Well-fertilized plants make more efficient use of water (in the sense of production of dry matter or marketable crop per unit volume of water used in evapotranspiration) than poorly nourished ones, an effect which has been recognized for many years.

This is partly because the more vigorous growth associated with fertilizer use can result in increased root development within the soil, so that soil water can be used to higher tensions and at greater depth. Adequate phosphorus shortens the development period and hastens maturity of the plant, thus increasing water-use efficiency by reducing the period of water loss by the crop through evapotranspiration. Potassium directly affects water loss by regulating the mechanism of opening and closing of the stomata and so reducing transpiration and respiration. Thus plants that are well supplied with potassium lose less water by transpiration.

Resistance to frost depends upon osmotic processes in the plant cells. These are influenced by the amounts of electrolytic substances developed (e.g. minerals, sugars, and soluble proteins), which function as a kind of "anti-freeze". Phosphorus and potassium, in particular, favour the storage of larger reserves of osmotic-effective compounds, which makes the plant less susceptible to frost and gives an increased chance of survival.

References

Beringer, H. and Trolldenier, G. (1979) Influence of K nutrition on the response to environmental stress. Proc. 11. Congr. Berne 1978, Internat. Potash Inst., 189–222.
Eifert, A. and Eifert, J. (1976) Relationships between K supply, yield of grapes, and frost resistance in vines. Potash Rev. (Berne) Subj. 29, 8. Suite, 1–6.
Gervy, R. (1970) Le phosphore et les effets du froid sur la végétation. Les phosphates et l'agriculture. DUNOD, Paris, 182–188.
Ignazi, J.C. (1977) Influence of climatic conditions on the response to phosphate in experiments on field crops. Phosphorus in Agriculture, No. 70, pp. 85–91.
Kresge, C.B. (1974) Effect of fertilization on winter hardiness of forages. In: Mays, D.A. (ed.) Forage fertilization. Amer. Soc. Agron., Madison, 437–453.
Larsen, J.B. (1978) Untersuchungen über die Bedeutung der K- und N-Versorgung für die Austrocknungsresistenz der Douglasie im Winter. Flora 167, 197–207.
Mussell, H. and Staples, R.C. (1979) Stress physiology in crop plants. J. Wiley and Sons, New York, pp. 510.
Steponkus, P.L. (1978) Cold hardiness and freezing injury of agronomic crops. Adv. Agron. 30, 51–98.
Turner, N.C. and Kramer, P.J. (Eds.) (1980) Adaptation of plants to water and high temperature stress. J. Wiley and Sons, New York, pp. 482.

E The effect of fertilizers on forage crops and animal health

E 1 Botanical composition of grassland

The botanical composition of grassland varies considerably according to geo-
graphical and topographical location, soil type, and climate. Fertile conditions, as in the lowland areas of temperate zones, suit the socalled "agricul-
turally-valuable" species such as perennial ryegrass (Lolium perenne), Italian
ryegrass (Lolium italicum), timothy (Phleum pratense), cocksfoot (Dactylis
glomerata), meadow fescue (Festuca pratensis) and tall fescue (Festuca
elatior). N fertilization favours these more aggressive and productive grasses.
Other generally less productive grasses also have a part to play in less
favourable conditions.

White and red clover are also important pasture plants and are favoured by
high available soil moisture capacity, mild climates, adequate supplies of
phosphate, potassium and lime, and essential minor elements. In upland areas
where intensive grassland production by N fertilization is not practicable
clover, other legumes and herbs are a very important source of N supply for
natural grassland.

When clovers are grown along with grasses they cannot compete very well,
especially in the presence of moderate to high levels of fertilizer nitrogen
which allow the more aggressive grasses to dominate the sward. In most
situations where fertilizer nitrogen is applied, even in the presence of ample
supplies of phosphate and potash, the percentage of clover in a mixed sward
is usually no more than about 5%, except in some favoured localities. In a
recent survey of permanent pasture in England and Wales it was found that
few fields had more than 5% clover (based on ground cover estimates).

Tall growing red clovers compete better with grasses than white clover but
they lack persistence, dying out usually in their second year. Intensive
fertilization up to economic levels generally has beneficial effects on sward
quality. Sward deterioration is more usually associated with mismanagement
such as occurs when pasture is either overgrazed or undergrazed or if animals
are allowed to "poach" the surface following heavy rain, or when cutting for
conservation is not carried out frequently enough or when the cut swath, for
ensiling or hay, is left on the ground too long. Bad distribution of slurry can
also seriously damage swards.

References

Charles, A.H. and Haggar, R.J. (Eds.) (1979) Changes in sward composition and productivity. Proc. Occasional Symp. No. 10, York, September 1978. British Grassland
Society. pp. 253.

FAO/ECE (1974) Symposium on the effects of fertilizers on the quality and nutritional value of grains, potatoes, selected fruits and vegetables, and forages, Geneva, June 1974.

Forbes, T.C., Dibb, C., Green, J.O., Hopkins, A. and Peel, S. (1980) Factors affecting the productivity of permanent grassland. A national farm survey. The Grassland Research Institute and Agricultural Advisory Service Joint Permanent Pasture Group, pp. 141.

Munro, J.M.M. and Davies, D.A. (1974) Potential pasture production in the uplands of Wales. 5 The nitrogen contribution by white clover. J. Brit. Grassl. Soc. 29, 213–223.

Penny, A., Widdowson, F.V. and Williams, R.J.B. (1980) An experiment begun in 1958 measuring effects of N, P and K fertilizers on yield and N, P and K contents of grass. 1. Effects during 1964–67. J. agric. Sci. Camb. 95, 575–582.

Koblet, R. (1979) Ueber den Bestandesaufbau und die Ertragsbildung in Dauerwiesen des Alpenraumes. Z. Acker- u. Pflzbau 148, 131–155.

E 2 The nutritional value of forage crops

The nutritional value of forage crops embraces their dry matter content, energy, crude protein, mineral composition and digestibility. N fertilizer reduces the dry matter percentage but, when applied in conjunction with adequate amounts of P, K and Mg fertilizers and lime, the yields of dry matter and energy and the content of minerals (N, P, K, Na, Ca, and some important trace elements) are increased substantially.

The percentage of total and digestible crude protein in forage are increased markedly by applications of N-fertilizers, and in swards with high clover contents. Percentage of non-protein nitrogen, which is of some value to the ruminant animal, is also increased by N fertilizer. Nitrate, which is part of the non-protein nitrogen fraction, has no nutritional value as such and may be harmful to the ruminant if present in very high concentrations. Under grazing conditions or normal systems of feeding silage, however, no risks are involved. More is said of this in Sections E 3 and E 4.

In most pastures, whether fertilized heavily or not, protein is present in excess of the requirement of ruminants. Any surplus nitrogen is recycled through dung and urine.

Clovers have higher protein and mineral contents than grasses and also tend to have a higher nutritional value; this is related to their higher digestibility. Too much reliance on clovers reduces total attainable yield of dry matter and if clover is present in excess this may result in nutritional upsets, such as bloat. Clovers have higher magnesium percentages in the dry matter than grasses but, as there is usually little active growth of clovers early in the spring, their limited presence does not materially help to prevent hypomagnesaemia. This condition can be effectively countered by applying Mg-containing fertilizers, feeding of calcined magnesite or dusting this material on the grass. In certain areas Mg and Na have to be supplied to pastures to supplement soils that have a deficiency of these elements.

Detailed studies have shown that even very heavy applications of N fertilizer to grassland generally do not have any adverse effect on the mineral composition of an all-grass sward.

References

Reith, J.W.S. (1980) Effect of nitrogen on the potassium requirement of herbage regularly cut for conservation. J. Sci. Fd. Agric. 31, 1076–77.

Saalbach, E. (1976) Mineraldüngung des Grünlandes in der Bundesrepublik Deutschland. Stikstof No. 83/84, 405–408.

Thompson, J.K. and Warren, R.W. (1979) Variations in composition of pasture herbage. Grass and Forage Sci. 34, 83–88.

Hartmans, J. (1975) The effect of intensification on crop composition and on health and production of livestock. Stikstof (Dutch Nitrogen Fertilizer Review) No. 18, 17–20.

Thomson, D.J. (1977) The role of legumes in improving the quality of forage diets. In: Brian Gilsenan (Ed.), Proc. Int. Mtg. on Animal Production from Temperate Grassland, 131–136. Irish Grassland and Animal Production Assoc.. An Foras Taluntais, Dublin, 1977.

Thöni, E. (1981) Die optimale Stickstoffdüngung der Gras-Weisklee-Mischungen. Mitt. Schweiz. Landw. 29, 161–171.

E 3   The yield and quality of animal products

By intensive use of nitrogenous fertilizers on grassland, in conjunction with adequate supplies of P, K, Mg and Na, the stocking density can be increased from 1−1.5 to about 3.0 cattle units per ha. In this situation grassland can provide practically all of the feed required during the grazing season and all of the roughage for the winter period. Thus, dependent on the production level of milk or meat and the management system employed, about 75−80% of the total feed requirements may be home-grown with consequent savings in feed costs compared with using expensive imported protein concentrates. At the same time, production of milk and meat is increased considerably. The results from many investigations have shown that, up to a fertilizer application level of 400 kg N per ha per year, one kg N applied as fertilizer will produce an additional 10 to 15 kg of milk or 1 kg liveweight gain.

Investigations have shown that rates of N fertilizer far in excess of economic optima have had no detrimental effect on milk yield per cow nor any appreciable effect on butterfat, protein or total solids percentages in the milk. The only discernible effect on milk quality was an increase in the non protein nitrogen fraction but this has no nutritional significance because the nitrate element of this, if present at all, is at a low enough level not to create any human health problems.

In German investigations, applying Mg-Kainit fertilizer to grass increased herbage intake with the result that milk yield increased by 2 litre per cow per day.

As mentioned in section E 1 legumes are favoured by P and K fertilization. Studies with cattle and sheep have in general confirmed that individual animal performance and milk yield have been higher on diets of legumes than of grasses. However, this effect on individual animal performance may have to be sacrificed to a certain extent in the attainment of optimum economic production of meat and milk on a hectare or farm basis.

References

Vanbelle, M. and Deswyeen, A. (1977) L'alimentation de la vache laitière haute productrice Rev. Agric. Brux. 30, 1487−1532.

Ernst, P. (1978) Einfluss der Magnesia-Kainitdüngung auf die Schmackhaftigkeit des Weidefutters und auf den Futterverzehr durch Milchkühe. Dissert. Inst. f. Pflz ernährung, Univ. Giessen, pp. 184.

Thompson, W. and Blair, T. (1981) The role of fertilizers in milk production. The Fertiliser Society London, Proc. No. 197.

Baker, H.K. (1981) The role of fertilizers in meat production. The Fertiliser Society London, Proc. No. 198.

Burg, P.F.J. van, Prins, W.H., Boer, D.J. den and Sluiman, W.J. (1981) Nitrogen and intensification of livestock farming in EEC countries. The Fertiliser Society London, Proc. No. 199.

E 4   Animal health and fertility

The so-called production diseases and metabolic disorders of dairy cows are caused by an imbalance or inadequacy of the necessary constituents, such as energy, protein, fibre and minerals that go to form a complete feed. High-yielding dairy cows are particularly prone to such problems and care has to be taken to feed them according to precise nutritional requirements and in relation to their level of production. Although fertilizer applied to grassland is sometimes incriminated when things go wrong there is little support for this in the scientific literature.

In recent years several very detailed and comprehensive, long-term experiments have produced evidence to show what effect very heavy dressings of fertilizer (much higher than the agronomic and economic optimum) have on the performance, health and fertility of dairy cows. Some of the findings from these studies are highlighted below.

Nitrate toxicity

Although the nitrate concentration in the dry matter of grasses rises with increasing amounts of N fertilizer the actual level is unpredictable because other factors such as light, temperature, moisture availability, age of plants, defoliation interval, etc. all affect the amount of nitrate that accumulates in the plant tissues.

In two separate long-term experiments in the Netherlands and the United Kingdom in which very high rates of 550 and 750 kg N per ha per year respectively were used on grassland, the substantial increase in the nitrate content of the grass dry matter that resulted from the higher N-rates did not deleteriously effect the performance and health of the dairy cows that were the test animals.

To cite one of these experiments as an example, the mean nitrate-N concentration of the herbage that received 750 kg N per ha per year over 5 years was 0.24% compared with 0.05% at 250 kg N per ha per year, with peak concentrations occasionally reaching 0.52%.

In a more recent grazing experiment in Holland 1000 kg N per ha per year was applied; nitrate-nitrogen concentrations up to 0.9% in the dry matter had no effect on methaemoglobin formation. The reason for this is that the ruminant is capable of consuming, without ill effects, considerable amounts of nitrate when spread over a 24 hour period as occurs in the normal process of grazing.

When confined to shorter intake periods, as in the feeding of hay, silage or turnips, however, the amount of nitrate that can be consumed without toxic symptoms developing is reduced to a lower level. Under normal feeding regimes, however, levels of nitrate in conserved forage are unlikely to cause problems.

Magnesium deficiency

The metabolic disorder known as hypomagnesaemic tetany is, amongst other factors, associated with a shortage of magnesium resulting from a low magnesium intake and/or reduced magnesium availability or resorption in the ruminant. The availability of magnesium to the cow is influenced to a large extent by the N and K contents of the diet. Other factors, such as low Na concentration and high NH_3 contents in the diet and low fodder intake, may also affect Mg-resorption in the digestive tract of the ruminant.

In intensive grassland management systems it is particularly important to ensure that adequate supplementation of magnesium is provided especially before and during the grazing of lush, heavily fertilized grass in early spring and in the autumn. As well as magnesium supplementation, the feeding of straw to grazing animals at such times can be a wise precaution.

In intensive grassland farming it is essential to maintain an optimum supply of nutrients. Nitrogen fertilization generally increases herbage Mg content, while potassium fertilization tends to decrease the Mg level in the herbage. However, nitrogen and potassium fertilization, by increasing the levels of N and K, crude protein and fatty acids in the herbage, may interfere with the resorption of Mg by the animal, thus increasing the tetany potential. Consequently more magnesium is required in the feed the greater the amounts of N and/or K that are ingested. Hypomagnesaemic tetany can be controlled by the following methods:

1. dusting pasture with magnesium oxide,
2. feeding magnesium-enriched concentrates, and
3. fertilizing with magnesium or magnesium-containing fertilizers.

Sodium deficiency

The daily requirement for maintenance and production of a dairy cow is 7 g Na and 0.5 g Na per kg milk. With a milk yield of 25–30 kg the Na requirement can adequately be met when the Na concentration in the herbage is 0.15–0.20%. Serious deficiency will occur with herbage containing 0.05% Na. The Na concentration in the herbage of intensively managed pastures can sometimes be too low. Fertilization can play an important role. If the soil has a low Na status sodium-containing fertilizers should be applied. If a potassium fertilizer is also required, a low grade K-fertilizer containing Na should be used.

If sufficient potassium is already supplied as animal manure, sodium should be applied as agricultural salt.

52

Copper deficiency

Copper deficiency has also been related directly to the intensive use of grassland. In the Netherlands cows grazing on high protein pastures tend to have a lower copper status than those grazing on low protein pasture. This agrees with reports that a decrease in the energy to protein ratio in the herbage has an adverse effect on the availability of dietary copper. However, in the UK experiment cited there was no difference between the blood copper levels in the herds grazing on low and high nitrogen pasture, although the levels in both herds decreased quite significantly during the trial. Copper deficiency can be corrected by feeding copper-enriched concentrates or by dressing of pasture with copper salts or injecting copper into the animal.

Fertility disorders

The use of high rates of fertilizers has sometimes been associated with general herd infertility problems. In fact, experimental evidence tends to deny any such association. Investigations carried out in Germany and Austria on non-specific herd sterility have shown that the fertility of animals is unrelated to the intensity of grassland management and level of fertilization. Results from investigations in the UK and the Netherlands indicate that the fertility of cows grazing high-N pasture was similar to that of cows on low-N pasture and compared favourably with the corresponding national data. Adequate supplies of nutrients in the total diet are essential to ensure high reproductive performance.

Investigations in the UK, Finland and the FR Germany with sheep and dairy cows indicated that high K content in the blood serum and high Na content in the saliva were correlated with high fertility.

It can be concluded that intensive and well-balanced use of fertilizers on grassland does not affect animal health or production and it is now recognised that many of these disorders can easily be prevented by employing appropriate husbandry measures.

References

Cheva-Isarakul, B. (1980) Zum Verhalten einiger Stoffwechsel- und Leistungsphysiologischer Parameter unter Nitratbelastung bei Milchkühen. Agrarwissenschaftliche Dissertation, Göttingen, pp. 111.

Coombe, N.B. and Hood, A.E.M. (1980) Fertilizer nitrogen: effects on dairy cow health and performance. Fertilizer Research 1, 157–176.

Martens, H. (1981) Neue Erkenntnisse über den Magnesiumstoffwechsel bei Wiederkäuern. Z. Tierphysiol. Tierernähr. Futtermittelkd. 45, 219–221.

McGregor, R.C. and Armstrong, D.G. (1979) The effect of increasing potassium intake on absorption of magnesium by sheep. Proc. Nutr. Soc. London 38, pp. 66.

Thalmann, A. et al. (1979) Nährstoff- und Mineralstoffgehalt von wirtschaftseigenen Futtermitteln aus Betrieben mit Fruchtbarkeitsstörungen. Wirtschaftseig. Futter 25, 133–146.

Küpper, M. (1976) Mengen- und Spurenelemente im Haar und im Blutserum von schwarz- bunten Rindern und ihre Beziehungen zu Leistungskriterien. Dissert. Univ. Bonn, Landw. Fak., pp. 116.

Rogers, P.A.M. (1979) Hypomagnesaemia and its clinical syndromes in cattle: A review. Irish Vet. J. 33, pp. 115.

Rowlands, G.J., Little, W. and Kitchenham, B.A. (1977) Relationships between blood composition and fertility in dairy cows – a field study. J. Dairy Res. 44, 1–7.

F Fertilizers in forestry

F 1 Wood quality

Fertilizers are used to improve tree growth and affect timber quality only indirectly through the increased growth; it has not been shown that fertilizers have any other influence on timber quality.

Any cultural measure, such as thinning or applying fertilizer, which is designed to improve growth will affect quality, but the magnitude of the change depends on the rate of growth before treatment. If, because of nutrient deficiency in the soil, the rate of growth is very low, the effect of increasing growth rate will be to increase the wood density. At normal rates of growth (annual ring width 1–2 mm) wood density may be reduced by 5–10%, while at higher rates the change in density is very small. If fertilizer does reduce density it still substantially increases dry matter yield. For example, a 30% increase in volume growth may correspond to 20% increase in dry matter production and a reduction in wood density of 5%.

So far as wood for *papermaking* is concerned, the rates of fertilizer used in practice produce only small changes in quality. There may be some loss in pulp yield through reduced wood density but this is partly compensated by an increase in thickness of the fibre cell walls of the early season wood: tearing strength of paper is maintained and tensile strength is increased. One advantage from using fertilizer is to produce a more uniform type of wood and this is a distinct advantage in papermaking.

As concerns timber for *building construction*, pit props etc., a decrease in density brought about by using fertilizer would somewhat reduce the strength of the timber, but there has been no case where rates of fertilizer currently used have caused any appreciable reduction in quality. On the other hand, there have been many examples where quality has been improved, the timber being more uniform and with fewer knots per unit volume, though the size of the knots may be increased by the use of fertilizer. Fertilizers as recommended and used in practice have been shown to have no effect on the shrinking characteristics of timber. For the production of veneer fertilized timber is preferred.

F 2 Fertilization and the environment

Fertilization increases the productivity of forest. It will change the natural environment but generally for the better.

In Europe and North America it is common practice to fertilize forests with nitrogen on mineral soils, and with phosphate and potash on sandy and peat soils. Single dressings of fertilizer are usually applied once in the life of a plantation; if more than once there is a gap of several years between applications. Although this practice introduces a risk of increasing the nitrate content of the watercourses, as a rule the effect is minimal and of short duration and does not present any measurable environmental hazard. The soil organisms and the wild animals living in the forest habitat are unharmed.

In acid forest soils urea seems to have a weak basic effect and in some cases this leads to a small degree of nitrification. Repeated fertilization with urea may therefore lead to the release of more nitrate into the groundwater. The nitrate content may also be increased when ammonium nitrate is used. Nevertheless the concentrations of nitrate found in groundwater have never exceeded the WHO health limits. The strongly reducing conditions that exist in the deeper layers of forest soils help to keep the nitrate content of groundwater at a low level.

On peat soils some phosphate is leached into watercourses from superphosphate but very little from rock phosphate.

It has not been established whether the combined effects of fertilization followed by clear felling years later have any detrimental effects on the environment. Nor is it certain that repeated fertilization following forest-clearing increases the amount of nitrate leached. Fertilization with urea could perhaps have such an effect because, more so than ammonia or nitrate, it seems to become incorporated in the organic matter of soil and so will tend to cause the release of more N by mineralization when the forest is felled. Up to three applications of N fertilizer during the life of a plantation, however, raises the total nitrogen content of the soil by an imperceptible amount.

One effect of fertilization is to change the vegetation of the forest floor from mosses and lichen to grasses and heather but this has no detrimental effects on the growth of the trees nor on the environment. Fertilization does not cause any visible change in the numbers or kinds of animals who use the forest as their habitat. Animals do not ingest significant quantities of the fertilizer granules when they are spread on the ground and no cases of toxic effects on forest birds or animals have been reported.

In some cases it has been observed that the trees acquire increased resistance against insect attack after fertilization but this can vary with locality and type of insects.

Just after fertilizers are applied the number of soil micro-organisms generally increases; with some species a brief toxicity effect is observed but very soon populations are back to normal again. Although fertilization generally

56

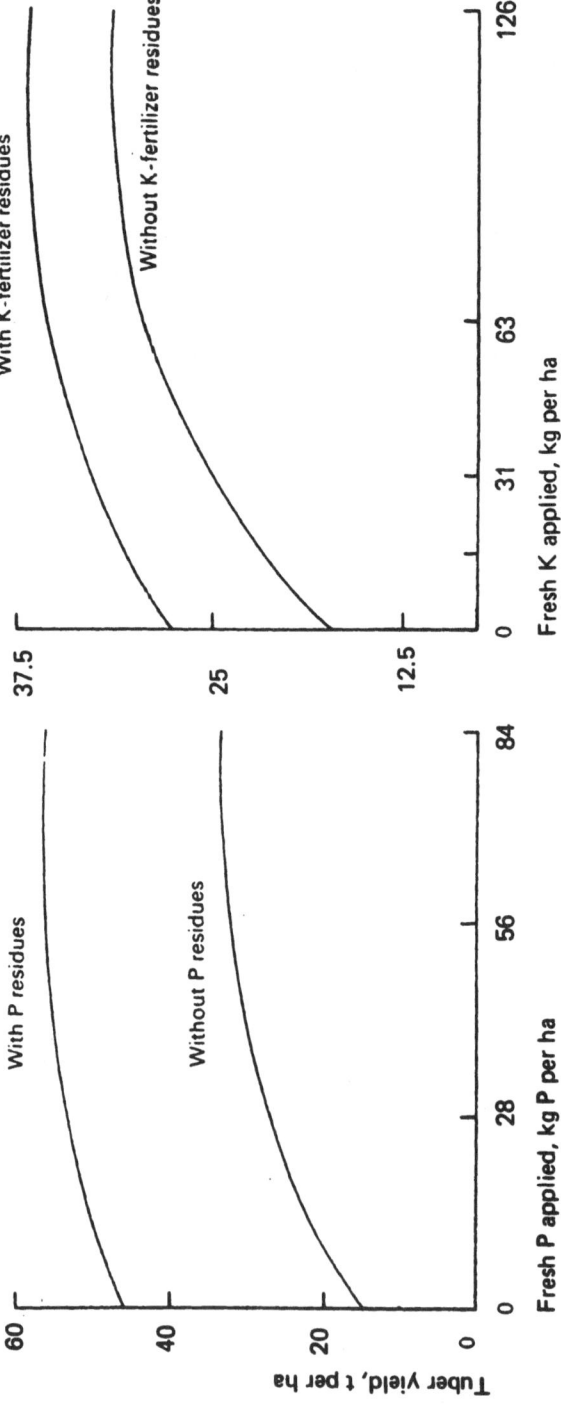

Figure G 1.1 Effect of fresh P and K applications on the yield of potatoes on soils enriched in P and K by previous manuring. From: G.W. Cooke (1967) The control of soil fertility. Crosby Lockwood and Son, London, pages 272 and 277.

has been found to increase the numbers of bacteria in the soil, the numbers of fungi tend to be reduced when urea is used.

References

Abrahamsen, G. (1970) Forest fertilization and the soil fauna. The Norw. J. of For. 78, 296–303 (section F 2).

Anderson, O. (1964) Forest fertilization, advantages and disadvantages (Norwegian) Från Sådd till Skord No. 10. Agro Tekniska Förlaget, Stockholm, Sweden.

Anonymous (1969) The influence of practical forest fertilization on the quality of the wood (Swedish). Institutet för Skogsförbättring, Information No. 1 (section F 1).

Baule, H. (1978) Wie wirkt sich die Düngung auf die Holzqualität aus? Hann. Landw. Forstw. Ztg. 131, 36–37.

Behrend, D.F. (1973) Wildlife management-forest fertilization relations. For. Fert. Sym. Proc. USDA For. Ser., Gen. Tech. report NE 3 1973, Upper Darby Pa. USA, 108–110 (section F 2).

Friberg, R. (1972) Nitrogen fertilization of forests (Swedish). Skogbrukets Informasjonsgrupp, Stockholm, 1972 (section F 2).

Gagnon, J.D. and Hunt, K. (1975) Effets de la fertilisation sur le poids spécifique et le rendement en pâte kraft du sapin baumier. Naturaliste Can., 102, 845–852 (section F 1).

Gladstone, T.W. and Gray, R.L. (1973) Effects of forest fertilization on wood quality. Forest Fertilization, USDA For. Serv., Gen. Tech. report NE 3, 167–173 (section F 1).

Klem, G.S. (1972) Forest fertilization and quality of timber (Norwegian). Skogeiren 1972, No. 3 (section F 1).

Moller, G. (1974) Practical and economic aspects of forest fertilization. Phosphorus in Agriculture, No. 62, 33–48.

Paavilainen, E. (1976) Effects of drainage and fertilization of peatlands on the environment. In: Tamm, C.O. (Ed.), Man and the boreal forest. Ecol. Bull. (Stockholm) 21, 137–141 (section F 2).

Salonen, K. (1980) Environmental effects of forest fertilization. Phosphorus in Agriculture, No. 77, 23–25.

Schulz, H. (1978) Einfluss der Düngung auf die Holzeigenschaften. Allg. Forstz. 15, 416–419 (section F 1).

Seibt, G. (1973) The effect of fertilization on the wood quality of pine and spruce. Symposium international FAO/IUFRO sur l'utilisation des engrais en forêt, Paris, 1973 (section F 1).

Van Lear, D.H. (1977) Growth and wood properties of longleaf pine following sylvicultural treatments. Soil Sci. Soc. Amer, J. 41, 989–995 (section F 1).

White, D.P. (1973) Effets des engrais sur la qualité du bois et d'autres produits forestiers. Symposium international FAO/IUFRO sur l'utilisation des engrais en forêt, Paris, 1973 (section F 1).

Wiklander, G. (1976) Forest fertilization and the environment (Swedish). Skogsägaren No. 10 (section F 2).

G Fertilizers and the soil

G 1   The effect of fertilizers on soil nutrient supply

Soil fertility, upon which plant growth and hence crop yield and quality depend, embraces its content of plant food (nutrients), its organic matter content, its structure, its ability to supply water and its depth. It has been said that fertilizers, though they stimulate crop yield, do nothing to bring about improvement in soil fertility. A few have even said that using fertilizers may cause soil fertility to deteriorate. There are no grounds for such opinions; in fact fertilizers, properly used, are most effective in improving soil fertility, not only as measured by nutrient content but also in terms of the other components of fertility. Further, such improvement is long-lasting.

The essential foods upon which plant growth depends are simple substances which occur in natural soils. The major nutrients (nitrogen, phosphorus, potassium, calcium, magnesium, sulphur) are taken up in large amounts. Obviously, the higher the soil's content of these nutrients in available form the higher is its fertility.

Because nutrients are taken up by the plant as ions, it makes no difference to the growing plant whether these ions originate from the soil itself, from organic manures or from fertilizers. When we apply fertilizer our aim is to increase the soil nutrient content. Returning farm wastes and crop residues to the land in the form of farm manures helps to *conserve* fertility but it does not *add* nutrients. When fertilizer is used, *extra* nutrients from outside the agricultural system are added to the soil and, thus, long-term soil fertility is built up.

If, as is often the case, the crop does not take up all of the nutrient added as fertilizer, the surplus is mainly retained in the soil so that when fertilizers are used regularly, soil nutrient reserves are built up. Excess nitrogen is easily lost from the soil by leaching and denitrification but N fertilizer does have positive long term beneficial effects on soil fertility (see Sections G 2 and G 3).

There is much evidence from long-term experiments to show that soil nutrient reserves built up by regular application of fertilizers are especially valuable. For instance Figure G 1.1 shows that, no matter how much fresh P or K fertilizer is used on potatoes, the yield on soils enriched in P and K by previous manuring is always above that on soil where the reserves have been depleted.

References

Banin, A. and Kafkafi, U. (Eds.) (1980) Agrochemicals in soils. Pergamon Press., Oxford, pp. 458.

Cooke, G.W. (1975) Fertilizing for maximum yield. Granada Publ. Ltd., St. Albans, pp. 297.

Drouineau, G. (Eds.) (1976) Very long-term fertilizer experiments. Annales Agronomiques 27, 483—1095.

FAO/UN (1979) Symposium on prospects of the use of fertilizers with a view to raising soil fertility and yields and of protecting the human environment. Geneva, 15—19.

Gutser, R. (1977) Möglichkeiten und Grenzen der Ertragssteigerungen in der pflanzlichen Produktion — aus der Sicht der Pflanzenernährung. Bayer. landw. Jb. 54, Sonderheft 1, 58—70.

Hoffmann, G. (1979) Düngen wir richtig im Blick auf die Bodenfruchtbarkeit, Ertrag und Gewinn? 90. VDLUFA Kongr. Augsburg 1978, Landw. Forsch. So.heft 35, 25—29.

International Potash Institue (Eds) (1980) K in the soil/plant root system. Research Topics No. 5, pp. 70.

Keller, E.R. (1980) Der Boden als Grundlage für die Erzeugung von Nahrungs- und Futtermitteln — Erhaltung seiner Ertragsfähigkeit auf lange Sicht. Schweiz. ldw. Forsch. 19, 209—222.

Mengel, K. and Kirkby, E.A. (1982) Principles of plant nutrition. Intern. Potash Inst., Bern, 3.ed., pp. 656.

Scheffer, F. and Schachtschabel, P. (1979) Lehrbuch der Bodenkunde. 10. Aufl., Enke-Verl., Stuttgart, pp. 394.

Tinker, P.B. (Ed.) (1980) Soils and agriculture. Blackwell Scientif. Publ., Oxford, pp. 159.

Tisdale, S.L. and Nelson, W.L. (1975) Soil fertility and fertilizers. 3. ed. Macmillan, New York, pp. 701.

G 2 The effect of fertilizers on soil organic matter

The term "soil organic matter" includes all organic materials while "humus" is used to describe finely divided non-living organic material without any residual structure. The organic matter of the soil makes an important contribution to fertility because it constitutes a reserve of slowly available nutrient (particularly nitrogen), stabilises soil structure, thus improving the habitat of the root, and increases its capacity to supply water.

Soil organic matter is formed by the biological and chemical breakdown of plant and animal residues in the soil. These residues include the roots of plants, stubble and other parts not removed at harvest, animal droppings, farmyard manure etc.

When fertilizers are used, the soil's supply of organic matter is increased because crop growth is improved, yielding more residues. Where animals are grazed on fertilized grassland, more stock can be carried to the hectare, hence there are more animal droppings and soil organic matter and nutrient levels build up. The direct contribution of green manures is increased through improved growth from fertilizer. Some data from long-term experiments in different parts of Europe serve to illustrate these points.

At Rothamsted (UK) on the continuous wheat experiment soil nitrogen contents were: after no manure or fertilizer 0.106%, after NPK fertilizer 0.121% and after farmyard manure 0.236%. At Askov (Denmark) on a light soil the corresponding figures after 50 years of a 4 course rotation, were: after no manure 0.106%, after fertilizer 0.118% and after manure 0.130%. At Grignon (France) soil receiving no manure or fertilizer for 77 years contained 0.118% N, soil receiving NPK fertilizer contained 0.142% N. Soils receiving all their nutrients as farmyard manure have the highest organic matter content, but the contribution of fertilizer via improved crop growth is considerable and, where the rotation includes a period under grass, it is very large.

References

Allison, F.E. (1973) Soil organic matter and its role in crop production. Elsevier, Amsterdam, pp. 637.

Dalal, R.C. (1977) Soil organic phosphorus. Adv. Agron. 29, 83–117.

FAO/UN (1975) Organic materials as fertilizers. Soils Bull. No. 27, pp. 400.

Hargitai, L., Myskow, W., Schwezowa, L. and Görlitz, H. (1981) Einfluss der Düngung auf den Humus- und Stickstoffgehalt der Boden. Intern. Z. Landw. 4, 362–365.

Kuntze, H., Voss, W. (1980) Statusbericht Düngung. Landwirtschaftsverl. Münster-Hiltrup, pp. 144.

Oberländer, H.E. (1979) Die Erhaltung des Humusgleichgewichtes in intensiv genutzten Ackerböden. Förderungsdienst 27, 16–19.

Russell, E.W. (1980) Soil Science: the last 50 years. In: Perspectives in world agriculture. Commonwealth Agric. Bureaux, Farnham Royal, 323–345.

Schnitzer, M. and Khan, S.U. (Eds.) (1978) Soil organic matter. Elsevier, Amsterdam, pp. 332.

Völker, U., Heisig, W. and Müller, G. (1980) Mehrjähriger Einfluss von Bearbeitungs-, Düngungs- und Fruchtfolgemassnahmen auf die Humusdynamik. Arch. Acker-Pflanzenb., Bodenkd. 24, 107–114.

G 3 Fertilizers and the soil flora and fauna

The soil flora and fauna play an important role in making a soil rich and fertile. Normal agricultural soils may contain up to 25,000 kg of living soil organisms per hectare. The quantity and activity of the soil micro flora and fauna are strongly influenced by the supply of transformable organic matter, pH (soil reaction), aeration, moisture content and cultivation of the soil.

Use of fertilizers increases the amount of organic matter available for the life processes of soil bacteria and fungi and so creates a more favourable environment in which these micro-organisms can develop. It also directly provides algae (like higher plants) with the nutrients they need for growth.

Suitable choice of fertilizers with an acidifying or liming effect can help to bring the pH of the soil within the optimum range for most crops (pH 6–7.5), which at the same time is the optimum range for most soil micro-organisms.

Trials comparing the number of soil micro-organisms in water-cultures and in plots either without fertilizer or with varying quantities of fertilizer nutrients (N, P_2O_5, K_2O) applied have shown that addition of all three nutrients, up to an optimum level, increases the population of micro-organisms. The optimum fertilizer level is of the same order as that for most crops. Even dressings which would be excessive for susceptible micro-organisms only temporarily (for no more than a month) depress their number and then by less than 10%. Afterwards the number increases to many times the original number. Calcium cyanamide, a nitrogen fertilizer with a side-effect on soil-borne fungal disease organisms, also only temporarily retards the activity of beneficial soil micro-organisms in the upper topsoil.

The larger soil organisms, such as earthworms, also flourish in well fertilized soils in which they are correspondingly well supplied with food.

References

Kuntze, H. and Voss, W. (1980) Statusbericht Düngung. Landwirtschaftsverl. Münster-Hiltrup, pp. 144.
Marshall, V.G. (1977) Effects of manures and fertilizers on soil fauna: A review. Spec. Publ. No. 3, Commonw. Bureau of Soils, Slough, pp. 79.
Martyniuk, S. and Wagner, G.H. (1978) Quantitative and qualitative examination of soil microflora associated with different management systems. Soil Sci. 125, 342–350.
Saive, R., Poelaert, J., Raimond, J., Brakel, J. and Huge, P.L. (1972) Influence sur la microflore du sol de diverses formulations d'engrais minéraux appliquès durant 62 ans. Rev. Ecol. Biol. Sol 9, 565–588.
Schmid, G. (1980) Biologische Aspekte beim Einsatz verschiedener pflanzenbaulicher Intensitätsstufen. Bodenkultur 31, 109–126.
Trolldenier, G. (1979) Effects of mineral nutrition of plants and soil oxygen on rhizosphere organisms. Proc. 4. Int. Symp. on Factors determining the behaviour of plant

pathogens in soil, München, 1978. In: Schippers, B. and Garns, W. (Eds.): Soil-Borne Plant Pathogens, Academic Press, London, 235–240.

Trolldenier, G. and Rheinbaben, W. von (1981) Root respiration and bacterial population of roots. I. Effect of nitrogen source, potassium nutrition and aeration of roots; II. Effect of nutrient deficiency. Z. Pflanzenernähr. Bodenkd. 144, 366–377, 378–384.

Van Vuurde, J.W.L. and De Lange, A. (1978) The rhizosphere microflora of wheat grown under controlled conditions. Plant and Soil 50, 447–474.

Walker, N. (ed.) (1975) Soil microbiology. Butterworth, London, pp. 262.

64

G 4 Fertilizers versus organic manures and wastes

The main function of fertilizers is to provide plants with nutrients in the forms and qualities best suited to their needs. For equal amounts of readily available plant nutrients, there is, in general, no difference in effectiveness between the nutrients from fertilizers and farmyard manure or other organic materials, since the latter are broken down by micro-organisms into the same inorganic ions as those contained in fertilizers before they are taken up by the plant roots. Organic manures have a low nutrient content and so have to be applied in large quantities. The bulk so applied influences the physical properties of the soil, but the availability of the nutrients it contains is uncertain. In comparison, using fertilizers gives greater flexibility in adapting the forms and quantities of nutrients to the plant's needs and has the further advantage that it involves handling a much smaller weight and bulk of material.

The slow release effect, which results from the organic bonding of the nutrients in farmyard manure and other organics applied to the soil, can vary substantially depending on soil moisture and temperature and on the activity of soil organisms.

High yields, resulting from adequate provision of all the nutrients needed by the plant, can be obtained either with fertilizers or organic manures alone, or with a combination of both. Very high yields in general need the well-timed application of readily available nutrients in the form of fertilizer.

There is a problem with organic waste materials, such as sludge and composted town refuse because in addition to useful nutrients, they may contain in some cases very high quantities of heavy metals (see Sections C 2 and J 7). Some trace elements (e.g. copper, zinc) are essential for plant growth but excessive amounts are harmful and may depress yield. The heavy metals have no nutrient function and can make food crops toxic if the content is high. There is also some risk in using these wastes regularly because these toxic materials may accumulate in soils to dangerously high levels. Careful control is needed.

References

Boguslawski, E. von and Debruck, J. (1972) Verwertung von Klärschlamm in Pflanzenbau. Mitt. DLG 11, 264–267.

Brogan, J.C. (Ed.) (1981) Nitrogen losses and surface run-off from landspreading of manures. Proc. Workshop in the EEC programme of coordination of research on effluents from livestock, Wexford, Ireland. Martinus Nijhoff/Dr. W. Junk Publishers, The Hague/Boston/London. pp. 471.

El-Bassam, N. and Tietjen, C. (1980) Flächenkompostierung kommunaler Abwasserschlämme Verfahren sowie Einfluss auf Ertragsbildung und Inhaltsstoffe von Böden, Wasser und Pflanzen. Landbauforschung Völkenrode 30, 51–78.

Farkasdi, G. and Glathe, H. (1973) Gefahr durch Müll-Klärschlamm-Komposte? Mitt. DLG 27, 771–772.

Garner, H.V. (1966) Experiments on the direct, cumulative and residual effects of town refuse, manures and sewage sludge at Rothamsted and other centres, 1940–47. J. agr. Sci. Camb. 67, 223–233.

Hucker, T.W.G. and Catroux, G. (Eds.) (1981) Phosphorus in sewage sludge and animal waste slurries. Reidel Publ. Co., Dordrecht, pp. 443.

Kiepe, H. (1972) Klärschlamm hilft Dünger sparen. Dt. Landw. Press 11, 5.

Loehr, R.C. (Ed.) (1977) Food, agriculture and agricultural residues. Proc. 1977 Cornell Agricultural Waste Management Conf. Ann Arbor Science Publishers, Ann Arbor. pp. 727.

Mays, D.A., Terman, G.L. and Duggan, J.L. (1973) Municipal compost: effects on crop yields and soil properties. J. Environ. Quality 2, 89–92.

Sticker, H. (1980) Schadstoffe im Bodem: Wann können Sie in die Nahrungskette gelangen? Schweiz. landw. Forschung 19, 267–279.

Voorburg, J.H. (Ed.) (1977) Utilisation of manure by land spreading. Seminar on utilisation of manures by landspreading, Modena, Italy. Commission European Communities. Publ. EUR 5672 e. pp. 732.

H Water Quality

H 1 Fate of fertilizer constituents

When fertilizer is applied to soil, the nutrients that it contains will in part be taken up by the crop, in part remain in the soil and in part be lost from the soil/crop system by one or more of a number of mechanisms. The relative amounts following these different pathways, and in particular the amount reaching surface or underground water, vary between nutrients and also with soil, climatic and agricultural circumstances.

Nitrogen. A high proportion of fertilizer nitrogen is taken up by the crop and a little is normally retained by the soil in organic form. Some may be lost by:
1. leaching to field drains or the underlying aquifer,
2. gaseous loss as nitrogen or oxides of nitrogen after denitrification, or
3. volatilisation as ammonia.
The relative amounts will vary very much according to conditions but an approximate estimate of representative apportionment for western European conditions is perhaps as shown in Table H 1.1.

Though both ammonium and nitrate forms of nitrogen are soluble, the former is strongly adsorbed on the soil colloids and consequently loss by leaching in this form is negligible. Loss only occurs in the nitrate form, so that soil and fertilizer are not sources of ammonium nitrogen in water. However, ammonium nitrogen applied to the soil in fertilizer is fairly rapidly nitrified, and once in the nitrate form is leachable.

Table H 1.1.

Process	% (range) of fertilizer
Crop removal in year of application	25−70
Leaching from fertilizer in year of application	0−30
Added to soil organic matter reserves and mineralised subsequently	0−30
Volatilisation as ammonia	0−40
Denitrification	0−30

The amount of leaching of fertilizer nitrogen will range from very little under normal agricultural conditions to appreciable amounts under unfavourable circumstances. The predominant factors are:

Factor	Less leaching	More leaching
Crop	Vigorous crop, Established crop, Grassland and other permanent crops	Poor crop or fallow, Seedbed application, Arable cropping
Soil	Heavy soil, Poor drainage	Light soil Good drainage
Time of N application	At the beginning of the main growing period or during active crop growth	At the end of or out of season, e.g. autumn or winter
Rate of N	At or below recommended rate	Above recommended rate
Climate	Low rainfall	High or irregularly distributed rainfall
	During period of soil moisture deficit	No soil moisture deficit

Phosphorus. Fertilizer phosphorus is poorly taken up by crops but that not taken up is effectively converted to water-insoluble form or is strongly adsorbed on soil particles. Loss to water systems by leaching is negligible. Where soil erosion is important, as on severely sloping fields under high rainfall, phosphorus adsorbed on soil particles may be carried into surface waters. For typical conditions, the result over a 3-year period could be as given in Table H 1.2 (though there will often be a continuing very slow crop removal over a much longer period).

Table H 1.2.

Process	% of fertilizer P	
	Average	Range
Crop removal (3-year period)	20	5–30
Held in soil	80	70–95
Leaching and run-off	less than 1	0–10

The amount of phosphorus leached will normally be low (less than 1 kg P per ha per annum) but may very occasionally be up to 2 kg P per ha. This greater loss may occur on very acid sandy soils, peat soils or poorly drained soils, or very rarely where massive P applications have saturated the P adsorption capacity of the soil, e.g., in some glasshouse situations.

Potassium. Fertilizer K is taken up by the crop to a very variable extent, the remainder being adsorbed on the cation exchange complex of the soil and subsoil, with very little lost by leaching. A typical sub-division is given in Table H 1.3.

Table H 1.3.

Process	% of fertilizer K	
	Average	Range
Crop removal in year of application	80–85	50–100
Held in soil and subsoil	15	0– 50
Leaching and run-off	less than 5	0– 10

Although potassium, together with other cations, is leachable, the amounts involved are normally small, less than 10 kg per ha per year. Cropping and fertilizer practice have little influence since K leached from the topsoil is normally retained in the subsoil and so not lost to plant roots. Leaching may be somewhat enhanced (up to 30 kg per ha) on sandy or organic soils of low sorption capacity and pH, or where excessive applications of slurry, etc. or fertilizer have raised the soil potassium status unduly.

Potassium entering the water system comes mainly from natural sources, i.e. from eroded potassium-bearing material and from weathering of the parent rock. Potassium in water does not present a health hazard, nor apparently a eutrophication problem.

H 2 Sources of nitrogen and phosphorus in water

Contribution from effluents. Fertilizers constitute only one source among many contributing to the nitrogen and phosphorus in natural waters. The first level of sub-divison to consider is between point sources of effluent (almost entirely sewage and industrial discharges) and diffuse sources, mainly land drainage. The importance of point sources is related to population density. With a mainly rural population of 1.5 persons per hectare about 30% of the nitrogen is from effluents and 70% from diffuse sources, while at 12 persons per hectare over 80% is from effluents. A much higher proportion of the phosphorus is derived from effluents − 75% at 1.5 persons per hectare and 99% at 12 persons per hectare.

Agricultural sources Diffuse sources of nutrient are mainly of agricultural origin, and include losses from livestock units, manure heaps, slurry lagoons and silage as well as drainage and surface run-off from farmland. It is difficult to apportion losses, but a Canadian investigation showed that highest nitrate levels in ground water were found close to houses and farmsteads.

Loss of nitrate from farmland includes nitrogen originally derived from all the sources that make up the overall supply to the crop − mineralized from soil organic matter, fixed by bacteria associated with leguminous plants or by free-living soil micro-organisms, added in the excreta of grazing livestock or in manures of animal origin, and in precipitation. Nitrogen fertilizer applied at the correct rate and time normally makes only a small contribution to the total.

The reasons that there is any loss at all from fertilizer are twofold:
1. for full crop production it is necessary to maintain a certain level of available nitrogen in the soil during the growing season. Rainfall is to an extent unpredictable and temporary excess may cause leaching,
2. crop residues containing nitrogen derived from fertilizer are partially mineralized at the end of the growing season and the nitrate so formed may be leached by autumn and winter rainfall.

The nutrient cycle in most agricultural soils is not − and cannot be − a closed one. The soil is continuously replenished by inputs from various sources and depleted by a number of routes, including leaching.

Negligible phosphorus is lost from agricultural soils or from fertilizer correctly applied to them.

Inorganic versus organic nitrogen It is sometimes suggested that the use of organic manures such as slurry or farmyard manure will reduce the risk of nitrate leaching. In fact, this is not so for a number of reasons:
1. they tend to be applied at very high rates, so that the residues left at the end of the growing season are liable to result in excessive amounts of nitrate being leached,

2. organic nitrogen is mineralised over a period of time and the phasing of this release does not coincide with the pattern of uptake by the crop,
3. organic manures are bulky and often can only be applied to land at times of year (e.g. autumn) when leaching loss can be expected.

As discussed in sections C 2 and G 4 the nutrients contained in organic manure are insufficient for the high level of agricultural production needed today and, in any case, most of them are already utilized.

Slow acting fertilizer. Slow acting nitrogen fertilizers release nitrogen over a period of weeks or months and are thus sometimes suggested as means of reducing leaching loss. Unfortunately, their N release is not readily predictable and may not supply enough available nitrogen during crop growth. Furthermore, mineralisation may continue beyond the growing period, increasing the amount of nitrogen at risk to subsequent leaching.

H 3 Effect of nutrient losses on water

Nutrient enrichment may affect water quality by causing problems associated with eutrophication or potability in relation to human health. These problems should be distinguished from pollution caused by direct contamination of surface waters by solid of liquid materials, toxic materials or pathogens.

Eutrophication. Eutrophication is the name given to the process by which waters are enriched with plant nutrients leading to enhanced growth of aquatic plants. It is a natural process, operating in nature on a geological timescale, but is accelerated by many aspects of human activity. Accelerated eutrophication, with excessive nutrients in a water body, can result in overgrowth of water plants and algae. This uncontrolled growth can result in deoxygenation, unpleasant smells, fish mortality and water treatment problems when the water is abstracted for use as a public supply.

Many aspects of man's activities have contributed to the eutrophication problems found in some areas in recent years — though it should be noted that eutrophication problems are not new and were encountered in some of the London water supply reservoirs before the end of the 19th century, when agriculture was relatively extensive and fertilizer use was negligible. Both industrial and agricultural activities can be responsible, but the prime factor is the rapid growth of human population and the introduction and extension of untreated or only partially treated sewage discharges.

Nitrogen and phosphorus are the key nutrients in accelerated eutrophication. The causation can be summarised as follows:
Upland and forest areas — N and P in water are both low and aquatic plant growth sparse.
Lowland agricultural areas — N in water moderate and P low, modest aquatic growth, limited by P levels.
Rivers and lakes receiving significant urban discharges — N and P in water are both elevated, risk of eutrophication.

The key feature in triggering eutrophication is thus usually the elevation of the P level as a result of sewage discharges. There may be exceptions, for example where P is lost from agricultural soils by erosion, or where large quantities of animal excreta are applied to the land or find their way into streams and rivers.

Fish population. Plant nutrients are essential for growth of aquatic plants, which provide food for freshwater fish, either directly or as the first stage in the food chain. Without nutrients there would be no fish and much research has shown that adding fertilizers to streams and lakes enhances fish production. Increases can be obtained by adding essential limiting elements, notably phosphorus and calcium.

Excessive quantities of certain nutrients may lead to eutrophication, with its ultimate stage of de-oxygenation when the plant material dies and decays. Deoxygenation is harmful to fish, particularly salmonids.

Nitrate and other nutrients from agricultural land have not been implicated in any of the spectacular cases of fish mortality that have been reported in the press. Excessive quantities of ammonia (e.g. from sewage effluent or animal manures) can be directly toxic to fish, though the ammonium ion is not. The lowest toxic concentration is 0.2 mg NH_3 per litre, for salmonids, but adverse effects may occur from 0.025 mg NH_3 per litre. These concentrations in leachate or runoff are not achieved as a result of properly conducted fertilizer application to land.

General experience suggests that fish mortality occurs, in practice, much more frequently from industrial incidents or sewage effluents than from agricultural causes and that, within the latter category, nutrient toxicity is likely to be of little consequence.

Nitrate and potability. Potability is the term used to describe the drinkability of water. An elevated nitrate content has certain medical implications (see Section J 1). Water supply authorities have a duty to ensure that nitrate content of public water supplies is below the World Health Organisation guidelines, and ensure that this is so by management or treatment of water supplies. Where nitrate is high fertilizer use is at most only a minor contributory factor (see Section H 2).

H 4 Nutrient contents of surface and ground water

Nitrogen (see also Section J 1). The total nitrogen contents found in rivers, lakes and underground water bodies can be from less than 1 mg per litre to about 10 mg per litre, and occasionally considerably higher. A certain mineral nitrate content must always be present because of the N in rainfall and as a result of fixation of N by certain blue-green algae. The factors affecting nitrate above this minimal level are:

Factor	Nitrate in water	
	Low	High
Annual rainfall	High	Low
Topography	Mountains & upland	Flat
Cropping	Forest	Arable (Grassland intermediate)
Livestock	Extensive	Intensive
Population	Sparse	Dense

Most rivers flowing through mountainous regions and forests contain less than 1 mg nitrate N per litre, rivers in grassland areas 1–5 mg per litre and rivers in arable areas 5–10 mg per litre. Occasional higher values are recorded in mainly agricultural rivers, due to short term "flushing" of nitrate from soils by autumn rainfall — these flushes may be particularly marked after a period of summer drought. Nitrate may be lost from slow-flowing rivers, lakes and water supply reservoirs as a result of uptake by aquatic plants or denitrification in bottom mud, so that large rivers and lakes tend to have considerably lower nitrate contents than ditches and small streams. There will also be a dilution effect in high rainfall areas.

Where nitrate nitrogen levels temporarily exceed 11.3 mg per litre (= 50 mg nitrate per litre) the WHO recommendation for drinking water in European countries, it is usual for the supply authority to store or to mix supplies, so that the limits are not exceeded. Where the nitrate level is continuously above this standard there is almost always a dominant effluent source of nitrate and the supply can only be used after treatment.

Most ground water contains 10 mg nitrate N per litre but some sources, e.g. in chalk or shallow sand aquifers, may be temporarily in excess of this. High nitrate levels are sometimes found in bands in the chalk aquifers of southern England, and are usually associated with loss of nitrate from the topsoil when mineralisation occurs after ploughing out grassland.

Phosphorus. The phosphorus content of streams is typically 0.01–0.1 mg per litre in forest watersheds and up to 0.5 mg in agricultural run-off. The latter usually contains a substantial proportion of P in particulate form, which may account for up to 80% of total P. The availability of this particulate P to aquatic organisms is often low, though this is difficult to quantify. Groundwater run-off (deep leachate) normally contains 0.02 mg P per litre. Water bodies receiving appreciable sewage effluents may contain considerably higher P contents; though much of this may be in the form of condensed phosphates, it will be fully available to aquatic organisms once hydrolysis to orthophosphate has taken place.

Potassium. Rivers and lakes normally contain less than 20 mg K per litre, derived from soil and rock weathering, and this provides no hazard (see Section J 4).

Trends in nutrient contents. In some rivers, especially in arable areas, there has been a slow increase in nitrate nitrogen content, usually at 0.1 mg per litre per annum, over a period of years. A major reason has been the intensification of agriculture so that greater crop residues give rise to more mineralised nitrogen, but growth of population and increase in sewage and thus in effluents is also important.

Nitrate levels in ground water have remained static in many areas, but have increased elsewhere; for example from 3.4 to 7.3 mg per litre in wells in the Netherlands over a 45 year period. Increases in nitrate in the chalk aquifers in southern England are due more to the change from extensive grassland to arable farming and to mineralization from ploughed out grass swards than to fertilizer use per se.

Phosphorus and potassium levels have changed little except where urbanisation has resulted in more point discharges.

References

Bottenberg, G. (1981) Auswirkungen der Stickstoffdüngung auf die Nitratbelastung des Grundwassers und Folgerungen für die Beratung. Forschung u. Beratung, MELF Nordrh.-Westf., Reihe C, No. 36, pp. 61.

Hebert, J. (1972) Influence de la fertilisation intensive sur la qualité des eaux superficielles et profonds. CIEC VIII Congres Mondial des Fertilisants, Vienna, 15–19 May 1972.

Henin, S. (1980) Rapport du Groupe de Travail activités agricoles et qualité des eaux. Ministère de l'Agriculture et Ministère de l'Environnement et du Cadre de la vie. p. 58.

Hood, A.E.M. (1976) Nitrogen, grassland and water quality in the United Kingdom. Outlook on Agriculture 8, 320–327.

Jürgens-Gschwind, S. and Jung, J. (1977) Ergebnisse von Lysimeteruntersuchungen in der Grossanlage Limburgerhof. BASF Mitt. für den Landbau, Heft 1, pp. 177.

Kolenbrander, G.J. (1973) Fertilisers, farming practice and water quality. The Fertiliser Society London, Proc. No. 135.

Ministry of Agriculture, Fisheries and Food (1976) Agriculture and Water Quality. Technical Bulletin 32. HMSO, London.

Royal Commission on Environmental Pollution (1979) Seventh Report. Agriculture and Pollution. HMSO, London.

Ryden, J.C., Syers, J.K. and Harris, R.F. (1973) Phosphorus in runoff and streams. Advances in Agronomy 25, 1-45.

Young, C.P. (1981) The distribution and movement of solutes derived from agricultural land in the principle aquifers of the United Kingdom, with particular reference to nitrate. Water Sci. Technology 13, 1137-1152.

Young, C.P. and Gray, E.M. (1978) Nitrate in groundwater; the distribution of nitrate in the chalk and triassic sandstone aquifers. Techn. Report TR 69, Water Research Centre, Medmenham, UK.

J Health aspects

J 1   Recommended safe levels of N in drinking water and the consequences of excess nitrate

World Health Organisation standards for European conditions are:

Recommended: less than 50 mg nitrate (11.3 mg nitrate N) per litre

Acceptable: 50–100 mg nitrate per litre

Unacceptable: over 100 mg nitrate per litre.

The maximum tolerance level for adults varies between 200 and 500 mg nitrate per litre depending on body weight, so high that there is practically no riks to adult health from using water supplied by a public authority. There is, however, a potential danger for babies up to 4 months of age (see Section J 2).

A study in the FR Germany showed that groundwater from catchments covered by woods and grass, not intensively grown and unmanured, generally contains less than 5 mg nitrate per litre. 1977 statistics on chemicals in water showed that in 96% of all samples of drinking water from wells the nitrate content was less than 50 mg nitrate per litre; 3.2% of the samples were close to the tolerance threshold of 90 mg nitrate per litre, and only 0.8% exceeded it. Contents in excess of 50 mg nitrate per litre were found mainly in water from wells that were not deep enough and therefore subject to contamination.

A Bavarian study concluded that on 98% of farms the practices employed do not contaminate the ground water. On average, the surface water from large water catchments in agricultural areas showed a content of 12.0 kg N per ha, while that from sources in generally unfertilized forested areas showed 12.9 kg N per ha.

When 180 kg per ha N as fertilizer is applied, and with 200 mm of drainage water per annum, lysimeter readings are as in Table J 1.1.

Table J 1.1.

Soil type	N in drainage water, kg per ha		
	Cereals	Root crops	Permanent pasture
Sandy soil	30	45	7
Loam and loess soils	21	32	5
Clay soils	15	24	3
Average	22	34	5

High nitrate concentrations in drinking water were as common in the past as they are today. Examples of high nitrate concentrations in well-water were reported from Leipzig 100 years ago. These were probably due to leakages from cesspools and sewage systems or to drainage from manure heaps. The cases of nitrate poisoning reported from the USA a few years ago were also the result of drinking water from private wells, while children who only drank tap water did not develop methaemoglobinaemia even when the nitrate content of the water was well above the recommended level. A group of leading American scientists is of the opinion that the intake of large amounts of nitrate by small children was only harmful in conjunction with gastroenteritis, an illness that may be spread by contaminated well-water. This agrees with the situation in the FR Germany, where nitrate concentrations of 50 mg per litre drinking water are regarded sufficiently safe for small children with regard to methaemoglobin formation.

References

Deutsche Forschungsgemeinschaft (1978) Chemische Wasserstatistik 1977.

Jürgens-Gschwind, S. and Jung, J. (1977) Ergebnisse von Lysimeteruntersuchungen in der Grossanlage Limburgerhof. BASF Mitt. für den Landbau, Heft 1, pp. 177.

Otto, A. (1980) Gewässerbelastung durch die Düngung in der Land- und Forstwirtschaft. Der Stickstoff No. 13, 40–52.

Schwille, F. (1978) Wasserwirtschaftliche Aspekte hoher Nitratgehalte im Grundwasser. Kolloquium Nitrat – Nitrit – Nitrosamine, Bochum, 9–10 February, 1–6.

WHO (1972) Water – Health hazards of the human environment, 59–71.

J 2   Methaemoglobinaemia in Europe

Methaemoglobinaemia is a metabolic disorder of the blood which, under special circumstances, can result from consuming food or water containing high nitrate or nitrite. It is rare in Europe even though the rates of N fertilizer used are among the highest in the world. There have been cases in children who had eaten spinach. Spinach takes up more nitrate than other plants. Under certain conditions of transport and storage this can be reduced to nitrite. In the home, nitrite starts forming in spinach prepared by the housewife if it is stored, particularly at high temperature, and when prepared in combination with milk. Repeated re-heating further promotes the process. It is, therefore, advisable to throw away left over cooked spinach. Low temperature as in the refrigerator delays reduction to nitrite but does not prevent it.

While nitrate in the amounts found in plants is not normally poisonous, nitrate can have a toxic effect if the haemoglobin metabolism is not yet fully functional, as is the case with infants up to about four months of age. In such infants, the methaemoglobin produced by the oxidation of haemoglobin in the presence of nitrite cannot be reduced again to haemoglobin since the infants are not yet able to produce the necessary enzyme, diaphorase. The oxidation of haemoglobin, or cyanosis, is dependent upon the action of nitrite (see Section J 8) and may be induced by the pathogenic bacteria which cause gastro-enteritis. Nitrogen fertilizers represent no danger to infants fed on spinach when the tolerance values for baby food of 250 mg nitrate per kg are observed, as required in the FR Germany.

References

Borneff, M. (1980) Untersuchungen an Säuglingen in Gegenden mit nitrathaltigem Trinkwasser. Zbl. Bakt. Hyg., I. Abt. Orig. 172, 59–66.

Breimer, T. (1982) Environmental factors and cultural measures affecting the nitrate content in spinach. Fertilizer Research 3, 191–294.

Molen, H. van der (1974) A test case against fertilizers in Illinois (USA). Stikstof (Engl. ed.) No. 17, –41–47.

Pufique, J. (1973) Fertilisation azotée et qualité des aliments en pots pour bébés (BBF). Cycle de Formation Permanente, Nancy, 10–12 April.

Philips, W.E.J. (1968) Changes in the nitrate and nitrite contents of fresh and processed spinach during storage. J. Agric. Food Chem. 16, 88–91.

Sattelmacher, P.G. (1963) Gefahren durch Nitrate im Trinkwasser? Fachblatt für Gastechnik und Gaswirtschaft sowie für Wasser und Abwasser 104, 134–135.

Swann, P.F. (1975) The toxicology of nitrate, nitrite and nitrosamines. J. Sci. Food Agric. 26, 381–382.

Temperli, A. (1981) Nitrat in unserer Nahrung. Chem. Rundschau Solothurn, 14 January.

J 3 Fertilizers, food crops and human health

Fertilizers increase crop yields and improve their nutritional quality (see also Section D). While plant composition, and, hence, nutritional quality is mainly determined by the genetic make-up of the plant, it is also influenced by many factors such as climate, soil, fertilizer usage, planting method, crop management, techniques of harvesting and conservation. Unlike organic manures, fertilizers can be applied in accordance with a plan which is matched to the needs of the plant. This leads to enhancement of nutritionally desirable constituents (protein, energy vitamins, minerals) and to a reduction of those that are less desirable (nitrate, oxalic acid etc.). For the detailed effect of individual nutrients see sections J 4–6.

Optimum quality is normally obtained at or near optimum yield. If fertilizer application is either well below or appreciably above the optimum, both yield and quality decline. The farmer will avoid excessive use of fertilizer in the interest of crop yield and quality and, above all, economy.

Good nutrition and human health demand agricultural products of the highest quality. Several experiments have shown that crops receiving fertilizers in the correct quantities have a higher nutritional value than those grown with organic manures alone. In a series of experiments with adults over three years, no differences in the general health or of medical test data were observed which could be related to the forms of fertilizers or manures applied to plants used in their diets (Wendt and co-workers).

References

Catel, W. (1950) Ueber den Einfluss der Verfütterung verschieden gedüngter Nahrungspflanzen auf das Gedeihen von Säuglingen. Landwirtsch. Forsch. 1, 53–69.

Fritz, D. (1978) Einfluss der Mineraldüngung auf die Qualität von Gemüse. Der Stickstoff No. 12, 14–22.

Glatzel, H. (1977) Sinn und Unsinn in der Diätetik. XII. "Biologisch" oder mineralgedüngt? Med. Welt 28, 253–311.

Schuphan, W. (1972) Effects of the applications of inorganic and organic manures on the market quality and on the biological value of agricultural products. Qual. Plant. Mater. 21, 381–398.

Wendt, H. and co-workers (1943) Ueber einen langfristigen Ernährungsversuch am Menschen mit verschieden gedüngten Gemüsen und Kartoffeln. Ernähr. 8, 281–295.

J 4   The effects of the primary nutrients N, P and K on human health

Nitrogen

Nitrogen is an essential element of plant and animal protein. The phosphorus-containing nucleo- and phospho-proteins are of particular importance in the metabolic processes. Enzymes and certain hormones are also composed of proteinaceous substances. Nitrogen, phosphorus and protein metabolisms are therefore closely interrelated. Nitrogen fertilizer applied in the right quantities and at the right times produces high yields of protein-rich fodder and food plants.

Phosphorus

The skeleton contains 75—80% of the total phosphorus in the human body and is important reserve of this element.

Phosphorus is also contained in proteins and lipids, e.g. in nerves, muscular tissue, internal organs, and in blood and milk. Adenosine triphosphate (ATP) is a vehicle for the transfer of chemical energy to the working muscle or of genetic information in the reproductive processes. Phosphorus deficiency in animals reduces the phosphate content of the blood serum, causing reduced fertility and bone fragility. Adequate amounts of calcium and phosphorus are important for the mineral metabolism of mammals. The calcium and phosphorus content of plants can be influenced by appropriate application of fertilizers, and their effects in improving the health and performance of sheep and rabbits has been demonstrated experimentally.

Potassium

As in the plant, potassium rarely occurs in organically bound form in animals and man, although it is abundant in the bone-forming cells. Potassium is important for the activity of muscle and nerve cells; it helps in the transport of sodium and, together with calcium, it has a regulatory effect on the heartbeat. Daily intake levels of adults are usually around 1.9—4.7 g K. Since 90% of the K intake is regularly excreted by the kidneys, a steady intake in the food is required. This need can be met by eating fruit and vegetables grown on soils receiving adequate amounts of potassium fertilizer and by drinking milk. In contrast to fruit and vegetables, the potassium content of cereal grains is little affected by fertilizer treatment. The potassium levels in milk and meat are relatively unaffected by the potassium content of the feed.

References

Evans, H.J. and Sorger, G.J. (1966) Role of mineral elements with emphasis on the univalent cations. Ann. Rev. Plant Phys. 17, 47–77.

Gadsby, D.C., Niedergerke, R. and Page, S. (1971) Do intracellular concentrations of potassium or sodium regulate the strength of the heart beat? Nature 232, 651–652.

Matrone, G., Smith, F.H., Weldon, V.B., Woodhouse, W.W., Peterson, W.J. and Beeson, K.C. (1954) Effects of phosphate fertilization on the nutritive value of soybean forage for sheep and rabbits. USDA Techn. Bulletin No. 1086, Washington, D.C.

Potash Inst. North America (1972) We are becoming more potassium conscious in human and animal nutrition. Better crops with plant food 56, No. 4, 2–10.

Reinberg, A. (1971) Biological rhythms of potassium metabolism. Internat. Kali-Institut Bern, 8. Colloq., Skokloster/Uppsala, 160–180.

82

J 5   Calcium, magnesium, sodium and sulphur in human health

Calcium

Calcium is the most abundant mineral in mammals and much is contained in
bone, blood and milk. Calcium deficiency, like shortage of phosphorus, can
cause bone damage (rickets, osteoporosis). No direct connections are reported
in the literature between human health and the application of calcium fertilizers to food plants, but the possibility cannot be ruled out that cows receiving fodder that is deficient in calcium (e.g., from pastures poor in Ca)
may give milk that is low in calcium, which in turn may cause bone damage in
children and too low a level of calcium in the blood serum.

Magnesium

Magnesium is essential for all vital metabolic processes in mammals, such as
the metabolic changes of sugar and protein or for the transmission of nervous
impulses. It is also a component of many enzymes, e.g., those involved in
carbohydrate metabolism. In combination with phosphorus and calcium,
magnesium regulates the colloidal state of the cell plasma, the osmotic
pressure of the blood fluids and the excitational state of the nervous system.
The ratio of magnesium to calcium in the cell is particularly important
(calcium – magnesium factor). Recent work has shown that magnesium also
plays an important role in the prevention of heart attacks. A normal, balanced
diet covers the maximum daily intake of 450 mg MgO required during
pregnancy and lactation. Magnesium deficiency is rare in man, but can arise as
a result of reduced resorption due to chronic diarrhoea or, in alcoholics, who
excrete much Mg in the urine. Tetany is the classic example of a deficiency
syndrome. Mg deficiency also causes collapses and disturbances in pregnancy.
Further confirmation is needed of reports suggesting that 5–10% of the adult
population suffers from magnesium deficiency because the MgO-requirement
increases as protein consumption increases. The increased incidence of
arteriosclerosis with magnesium deficiency, as observed in animal experiments,
cannot be directly related to humans, but doctors and nutritional experts
agree that more attention must be paid to the value for human health of
magnesium-rich soils and food.

Sodium

In humans sodium deficiency is mostly of transitory nature, e.g., after pro-
fuse sweating, when the body loses more sodium than can be replaced from
the food. In this case the remedy is to take salt tablets. Excessive intake of
sodium causes high blood pressure and danger of collapse, but potassium,
which is simultaneously ingested with many different foods and which acts

antagonistically to sodium in metabolic processes and in muscle, serves as a protective "antidote".

Sulphur

Sulphur is an essential consituent of many amino acids and of the proteins that are formed from them.

References

Fleischel, H. (1970) Düngung und Tiergesundheit. Verlag Gerhard Rautenberg, Leer/ Ostfriesland, pp. 72.
Kurmies, B. (1957) Ueber den Schwefelhaushalt des Bodens. Die Phosphorsäure 17, 258–278.
Lang, K. (1970) Biochemie der Ernährung. 2. Aufl., Verlag Dietrich Steinkopff, Darm- stadt, pp. 694.
Mengel, K. (1972) Ernährung und Stoffwechsel der Pflanze. Verlag Gustav Fischer, Stuttgart. pp. 470.

J 6   Micronutrients and human health

Many micronutrients are present in the human body but few of them are known to function in cellular metabolism. Only iron and cobalt appear to cause dietary problems.

Iron

Iron is the central atom of haemoglobin, the respiratory pigment of man and other mammals. Iron is also a constituent of many enzymes. Iron deficiency leads to various disturbances, particularly anaemia and deficiency of respiratory enzymes, which affects cell growth. A normal diet supplies enough iron for man but anaemia due to iron deficiency has increased among women of child-bearing age in Western Europe and the USA. This may be due to the relatively low iron content of industrially processed foods and the use of iron-free, non-rusting kitchen utensils. In the USA an attempt is being made to combat this by Fe enrichment of flour and other cereal products. Such an enrichment should also have a favourable effect during influenza epidemics, since infections reduce the level of iron in the blood serum. There are no data on possible direct effects of the use of iron fertilizer on health.

Cobalt

As a constituent of vitamin B12, which controls protein metabolism and haemoglobin synthesis, cobalt is required by both man and animals.

Copper

Copper plays a part in haemoglobin synthesis and is stored in the liver. For humans with a balanced diet there appears to be no danger of copper deficiency. No precise statement can be made, however, concerning possible consequences of deficiency or excess of this element in food or of the effect of fertilizer use.

Manganese

The highest concentration of manganese in mammals is in bone. Mucous membranes, liver, pancreas and gastro-intestinal tissues also contain relatively high concentrations. Manganese is essential for important metabolic processes, particularly for reproductive functions. It is unlikely that dietary manganese deficiency occurs in humans. Excess of manganese causes poisoning, with symptoms similar to those of viral meningitis and has so far only been observed in workers in the manganese mines after excessive inhalation of ore dust. It is extremely unlikely, therefore, that the presence or absence of manganese in fertilizers is of any significance for human health.

References

Balks, R. (1955) Mangan in der Tierernährung. Die Phosphorsäure 15, 162–171.

El-Bassam, N. (1978) Spurenelemente: Nährstoffe und Gift zugleich. Kali-Briefe 14, 255–272.

Fleischel, H. (1970) Düngung und Tiergesundheit. Verlag Gerhard Rautenberg, Leer/Ostfriesland, pp. 72.

Lisk, D.J. (1972) Trace metals in soils, plants and animals. Adv. Agron. 24, 267–325.

Riehm, H., Schultze-Grobleben, W. and Baron, H. (1954) Die Mikronährstoffe Bor, Kupfer und Kobalt im Thomasphosphat und ihre Bedeutung zur Verhütung von Mangelkrankheiten. Die Phosphorsäure 14, 55–68.

Vahrenkamp, H. (1973) Metalle in Lebensprozessen. Chemie in unserer Zeit 7, 99.

J 7   Heavy metals and human health

From time to time concern is expressed about the possibility of toxic heavy metals entering the food chain. These heavy metals occur all over the world mainly in deposits and waste tips from mining operations but they do occur in traces in natural soils. They enter the environment through:
The burning of fossil fuels
Spreading of town and industrial wastes
Accidental discharge from factories
Fertilization.
Some fertilizers may contain very small amounts of lead and small but variable amounts of cadmium and only minute traces of mercury.

Lead

In the old days people were much more exposed to lead (from water pipes, glaze on pottery, pewter vessels and paints) than is the case today when the main pollutant is high-octane motor-fuel. The degree of pollution that might be occasioned through fertilizer use is quite insignificant in comparison with the other sources of lead.

Cadmium (Cd)

The average soil contains about 0.5 mg Cd per kg, equivalent to 1500 g Cd per ha. In rock minerals from which phosphatic fertilizers are manufactured cadmium, too, occurs as a natural impurity amounting to 1–90 mg Cd per kg. Fertilizer manufacturers calculate that at the present status of technology about 80% of this impurity remains in the fertilizer. As a result of using fertilizers manufactured from the 13.5 million t phosphate rock imported into the EEC in the year 1980/81 less than 60 g Cd per t of P_2O_5 (range 12–100) were applied to European farms. Thus an average application of 50 kg P_2O_5 per ha adds from 0.6–5 g Cd per ha to the soil, in the average about 3 g. Compared with the average Cd content of the soil this amount added by fertilizer can make no practical difference. Furthermore, the greater part of fertilizer cadmium is quickly transformed in the soil into forms which are not easily taken up by the plant.

The amounts of cadmium added to the soil by phosphate fertilizers are low in relation to the amounts permitted by regulations pertaining to the agricultural use of sludges and industrial wastes, e.g.:

15 g per ha and year in Denmark and Sweden
20 g per ha and year in Finland
25 g per ha and year in the FR Germany
30 g per ha and year in Norway
60 g per ha and year in France
167 g per ha and year in the United Kingdom.

It is estimated that appreciable amounts of cadmium are added in rain and atmospheric dust: about 3 g per ha and year in Denmark, about 6 g in the FR Germany and up to 19 g per ha in Belgium. The permitted level in industrial areas of the FR Germany is 27 g per ha and year.

To conclude, it can be said that the amounts of cadmium in phosphatic fertilizers do not constitute environmental risk, especially when considered in relation to other sources of pollution. The risk of enhanced cadmium uptake by crops can be greatly reduced by correction of soil acidity through liming.

Urban wastes can be useful in appropriate situations but it should be pointed out that the use of fertilizers carries with it far less risk of pollution than does the use of such wastes, which may, according to source, be strongly contaminated with heavy metals. There are other risks in using sewage sludge and composted waste, which may contain carcinogenic polycyclic hydrocarbons and pathogenic organisms, risks that do not occur when fertilizers are used (see also Sections C 2 and G 4).

References

Bundesverband der Deutschen Industrie (1981) Stellungnahme des Bundesverbandes der Deutschen Industrie zum Bericht des Umweltbundesamtes "Ein Beitrag zum Problem der Umweltbelastung durch nicht- oder schwer abbaubare − Stoffe dargestellt am Beispiel Cadmium". Bundesverband der Deutschen Industrie e.V., Köln, 1−28.

Davis, R.D. and Coker, E.G. (1980) Cadmium in agriculture, with special reference to the utilisation of sewage sludge on land. Techn. Report TR 139, Water Research Centre, Medmenham, UK. pp. 112.

Diehl, J.F. (1981) Die Belastung des Verbrauchers durch Cadmium − eine kritische Übersicht. BFE-Bericht Nr. 3, 1−88.

Diehl, J.F. (1981) Schwermetalle in Lebensmitteln. Vortrag VCI-Pressegespräch Frankfurt, 19 May.

Herms, U. and Brummer, G. (1980) Einfluss der Bodenreaktion auf Löslichkeit und tolerierbare Gesamtgehalte an Nickel, Kupfer, Zink, Cadmium und Blei in Böden und komposierten Siedlungsabfällen. Z. Pflanzenernähr. Bodenkunde.

Isermann, K. (1981) Einflüsse Cd-haltiger Düngerphosphate auf den Cadmiumgehalt landwirtschaftlich und gärtnerisch genutzter Böden. Vortrag Cadmium-Anhörung vor Bundesministerium, Bundesgesundheitsamt und Bundesumweltambt Berlin, 2−4 November, 1−7.

Kampe, W. (1981) Die Fremdstoffsituation im Gesamtverzeher von Obst, Gemüse und Kartoffeln. Kali-Briefe 15. 635−646.

Kick, H. and Poletschny, H. (1978) Ein Kurzbericht über langjährige Feldversuche mit Müllkomposten und Klärschlämmen. Schwermetallgehalte in der Erntemasse. Landwirtsch. Forsch., So.heft 35, 412–418.

Kloke, A. (1981) Aufnahme umweltrelevanter Elemente durch die Pflanze. Vortrag Tagung der Deutschen Gesellschaft für Qualitätsforschung e.V., Speyer; 26, 1–16.

Mineral elements'80 (1981) Proc. Nordic Symposium on soil-plant-animal and man, interrelationships and implications to human health. 9–10 December 1980. Helsinki. Part I, pp. 328 and Part II, pp. 639.

Niagru, J.O. (Ed.) (1980) Cadmium in the environment. Part I Ecological cycling. J. Wiley and Sons, New York/Chichester/Brisbane/Toronto/Singapore. pp. 682.

Niagru, J.O. (Ed.) (1980) Cadmium in the environment. Part 11: Health effects. J. Wiley and Sons, New York/Chicester/Brisbane/Toronto/Singapore. pp. 908.

Pluquet, E. and Feige, W. (1979) Tolerierbare Schwermetallgehalte von Abwasserfaulschlamm in Abhängigkeit variabler Bodeneigenschaften und Pflanzenverträglichkeit. Jahresbericht 1978 zum Forschungsvorhaben 10304051 an das Umweltbundesambt Berlin.

Seybold (1981) Bericht 3rd Intern. Cadmium-Konferenz in Miami, 3–5 February, 1–5.

J 8 Nitrosamines and nitrogen fertilizer

Carcinogenic nitrosamines are formed by the interaction of nitroso compounds and alkylamines. Much work has been done to determine whether, and how, nitrate in feed and foodstuffs may be involved in nitrosamine formation. Some nitrosable compounds will always be formed in living organisms in the course of nitrogen metabolism. When these come into contact with nitrite, nitrosamines may be formed. The use of nitrogen fertilizers may increase the nitrate content of plants or plant organs. Nitrate itself is not involved but in certain conditions it can be reduced to nitrite in animals and man thus providing one of the raw materials for the synthesis of nitrosamines. This applies especially in the absence of inhibitory substances like vitamin C.

Nitrosamine formation depends on many factors such as acid conditions, warmth, microbial activity and catalysts like thiocyanate, iodine, bromine, chlorine and formaldehyde. Nitrosamines have also been found in certain milk products and in smoked meat and fish. The addition of black pepper and paprika can promote nitrosamine synthesis. Some organic acids, for example ascorbic acid, that occur frequently in nature, inhibit their formation so that eating adequate fruit and vegetables offers some protection against cancer of nitrosamine origin. Analysis of samples from differently fertilized plants has shown that, as far as can be proven by present techniques, nitrosamines are not formed in crops whether or not they receive fertilizer. Plants can take up nitrosamines through their roots and can use them to some extent as a source of nitrogen, but they do not accumulate nitrosamines. Although half decayed spinach and beetroot contained substantial amounts of nitrate and nitrite, no nitrosamines were detected.

In animal experiments it was impossible to influence the level of nitrosamines in the blood either by increasing the nitrogen content of the feed or by varying the source of nitrogen.

It can be said that there is no proven direct connection between the use of nitrogen fertilizers and the formation of nitrosamines in feed and foodstuffs. To quote the Royal Commission on Environmental Pollution, "... there is no evidence that unambiguously associates nitrates and N-nitroso compounds in human tissues or body fluids with certain carcinoma of any organ in man".

References

Dressel, J. (1979) Das nitrosaminproblem – Zusammenhänge mit der Stickstoffdüngung. Vortrag Symposium "Nitrosamine" der Deutsche Forsch. Gemeinsch., Bochum, 1–15.

Dam Kofoed, A., Nemming, O., Brunfeldt, K., Nebelin, E. and Thomsen, J. (1981) Investigations on the occurrence of nitrosamines in some agricultural and horticultural products. Acta Agriculturae Scandinavica 31, 40–48.

Eisenbrand, G. (1981) N-Nitrosoverbindungen in Nahrung und Umwelt. Wissensch. Vortragsges. mbH Stuttgart. 1–134.

Kampe, W. (1981) Stickstoffdüngung und Gesundheit. AID-Verbrauchersdienst 26, 36–38.

Kupfer, H. (1978) Düngung und Ertrag von Gemüse – die Entwicklung seit 1958. Gemüse 14, 284–286.

Royal Commission on environmental pollution (1979) Seventh Rep. Agriculture and pollution. HMSO, London, Cmnd. 7644.

Temperli, A., Künsch, U. and Schärer, H. (1979) Untersuchungen über den Nitratgehalt in industriell hergestellter Säuglingsfertignahrung und Gemüsesäften. Schweiz. landw. Forschung 18, 33–36.

J 9 Nitrogen fertilizers and stratospheric ozone

Concern has been expressed in recent years that various areas of human activity could lead to gross reduction in the amount of ozone in the stratosphere with consequent danger to the life or health of humans, animals and plants.

It has been mooted that the production of oxides of nitrogen by denitrification in the soil or in surface waters of nitrate originating from fertilizer and plant and animal remains could contribute to the possible reduction in stratospheric ozone. Other sources of nitrogen oxides are their production by supersonic aircraft in the stratosphere and the testing of nuclear weapons. Halogenated hydrocarbons used as aerosol propellants might also be implicated.

It is not disputed that there are such effects but there are widely varying estimates of the magnitude of the effects in relation to the total amount of ozone present. There are major gaps in knowledge and a lack of appropriate data that would enable meaningful predictions about such effects to be made. Some early accounts of the possible effects of nitrogen fertilizer are now considered to have been much exaggerated and, despite the dire early prognostications, so far any reduction in stratospheric ozone has been small and of doubtful significance. Separate attempts at estimating the effect of increased denitrification on ozone have included a decrease of 20–30% by the year 2025 and decreases of less than 10%, 4% and 1 to 4%. More recent work, based on new rate coefficient data predict small increases in ozone rather than a decrease.

In view of the many uncertainties, the possibility of a significant decrease can only be a matter of conjecture and should not influence local or global policies on fertilizer use at this time.

References

Anonymous (1979) Chlorofluorocarbons and their effect on stratospheric ozone. 2nd. Rpt. Pollution Paper No. 15. Department of the Environment, Control Directorate on Environmental Pollution. HMSO, London.

Crutzen, P.J. and Enhalt, D.H. (1977) Effects of nitrogen fertilizers and combustion on the stratospheric ozone layer. Ambio 6, 112–117.

Liu, S.C., Cicerone, R.J., Donahue, T.M. and Chameides, W.L. (1976) Limitation of fertilizer-induced ozone reduction by the long lifetime of the reservoir of fixed nitrogen. Geophysical Research Letters 3, 157–160.

McElroy, M.B., Elkinds, J.W., Wofsy, S.C. and Yu (1976) Sources and sinks for atmospheric N_2O. Reviews of Geophysics and Space Physics 14, 143–150.

Mineral elements'80 (1981) Proc. Nordic Symposium on soil-plant-animal and man, interrelationships and implications to human health. 9–10 December 1980. Helsinki. Part I, pp. 328 and Part II, pp. 639.

Timmerman, F. (1979) Stickstoffdüngung und Ozon in der Stratosphäre. Der Stickstoff No. 12, 71–81.

Turco, R.P., Whitten, R.C., Popoff, I.C. and Capone, L.A. (1978) SSTs, nitrogen fertilizer and stratospheric ozone. Nature, London 276, 805–807.

Whitten, R.C., Boruck, W.J., Capone, L.A., Riegel, C.A. and Turco, R.P., (1980) Nitrogen fertilizer and stratospheric ozone: latitudinal effects. Nature 283, 191–192.

K Energy

K 1 Energy and agriculture

Energy inputs into agriculture may be freely available (e.g. sun, wind), renewable (e.g. organic matter, human and animal power), finite (e.g. fossil fuels) and, in some cases, from developing nuclear sources. Traditional agriculture relies mainly on freely available and renewable energy sources, but more intensive agricultural systems consume increasing quantities of finite fossil fuels. As a result, land and declining labour resources are more productive. Western civilisation requires intensive food production systems to feed large urban populations and, in some cases, for export to less well favoured nations.

Even so, the amount of energy consumed in agriculture, as a proportion of the total energy used by some developed countries, is low – about 3.5% globally; with an estimated range of about 3% (FR Germany), 4% (United Kingdom) to 8.5% (France), while a major food exporting country such as the USA only uses 2.8% of its total energy consumption in agricultural production.

Extensive agricultural systems may produce 2 t of dry matter per hectare, while intensive agricultural systems can produce about 20 t of dry matter per hectare. However, intensive agricultural systems use more finite sources of energy to achieve this tenfold increase in production. This energy may be in the form of diesel fuel to drive agricultural machinery or dry crops, as fertilizers to provide adequate nutrients for plant growth, as chemicals to reduce losses due to pests and diseases and even indirectly as buildings to store production. An example of the various forms of primary energy used in intensive agricultural production is given in the following Table K 1.1.

Table K 1.1. *Percentage primary energy consumption.*

Item	%
Power – fuel, oil, electricity	38
Machinery	15
Fertilizers	33
Agrochemicals	2
Buildings	5
Miscellaneous	7
Total	100

The amount of energy used in agriculture is only part of the equation. Consideration must also be given to the human food energy output. In the case of most arable crops, for every unit of energy input 2 to 3 units of food energy are produced, reflecting the plant's ability to capture the free energy of the sun.

In the case of livestock, plants and/or grain are consumed to produce human food (milk and meat). Animals are not such efficient converters of energy into livestock production, only 0.2 to 0.4 units of human food energy are produced. But animal products are valued not only as sources of human food energy, but as sources of protein, vitamins and minerals and contribute to the perceived standard of living. Thus strict energy analysis in human food energy terms alone does not value the contribution of agriculture to the production of other essential human nutrients, clothing, housing, industrial raw materials, the mode, standard and quality of life.

Even so, agriculture uses energy comparatively efficiently when compared with other industries. Strict total energy input: output ratios for intensive national agricultural industries range from 1 : 0.5 to 1 : 0.7, while those of many other industries are 1 : 0.2 to 1 : 0.4.

A reduction of agricultural energy use in developed countries, say of 10%, would only have a minor effect on total energy consumption (0.2 to 0.8%), but a major effect on total food production. For this reason it is very important to give due consideration to the role of energy in agriculture and its fundamental contribution to the maintenance and development of mankind.

K 2 Fertilizer energy

Intensive agricultural systems rely on increasing quantities of fertilizers to increase the productivity of land and labour. Fertilizers are used to correct the nutrient deficiencies of the soil and thus to provide the nutrient requirements of crops. Adequate nutrition is an essential condition for healthy plant growth and the achievement of satisfactory crop yields. The level of yield also directly reflects the ability of that crop to capture the free energy from the sun. Directly and indirectly fertilizers represent another form of "energy" for plant growth and yield. Fertilizers make a very positive contribution to the ten-fold increase in yield from intensive agricultural systems compared with that achieved from extensive agriculture.

Most countries rely on organic fertilizers (plant, animal and human residues) to some extent; as agricultural intensity increases, more fertilizers are applied as inorganic chemicals. This practice enables a more appropriate balanced range of plant food nutrients to be readily available for healthy plant growth, reduces the possible hazard of disease from organic residues and is usually more cost effective.

Energy is used to mine, extract, transport, manufacture, process, store and distribute fertilizers. Of the major plant food nutrients, phosphate (P_2O_5) and potash (K_2O) require little energy to convert them into useful fertilizers. Nitrogen (N), the major nutrient of plants, while freely available in the air, has to undergo a major chemical process to convert it into a fertilizer. Thus the energy requirement in the manufacture, distribution and application of fertilizers is in the order of:

N	60–80	MJ per kg N
P_2O_5	10–20	MJ per kg P_2O_5
K_2O	5–10	MJ per kg K_2O

The above figures are representative of the total energy requirement of the major inorganic plant food nutrients, including extraction, manufacture, packaging, storage and distribution in developed agricultural economies. However, where fertilizers are transported over increasing distances the required input for distribution increases, and is only partially offset by increased concentration of plant food nutrients per unit weight of product.

A reasonable agricultural soil without any applied manure may produce about 1 t of grain per hectare. With organic manure the grain yield can be doubled to 2 t per ha. In the same situation the application of a balanced inorganic fertilizer of 140 N : 80 P_2O_5 : 80 K_2O kg per ha will increase the yield to 6 t per ha. Because inorganic nitrogen represents the major fertilizer energy input in crop production, and is also the major promotor of plant growth, the following example of temperate cereal production illustrates the energy input: human food energy output benefit:

Grain yield – Nil N	2.3 t per ha
– 200 kg N per ha	5.9 t per ha
– N response	3.6 t per ha
Kg grain per kg N	18.0
MJ grain human food energy per kg N	280
Ratio N energy: human food energy	1 : 4 approximately

Also the straw produced in grain production has an energy value which has not been included in the given example.

While inorganic fertilizers are often the largest single source of energy input in arable crop production, they represent about 30% of the total energy input in intensive farming systems. Thus the elimination of inorganic fertilizers from intensive crop production might reduce the total energy input by one third, resulting, however, in less than half the potential grain yield and a substantial decline in the total energy input: human food energy output ratio.

It is essential that the role of inorganic fertilizers in agriculture is kept in perspective. Their use increases the plant's ability to capture free energy from the sun promoting healthy growth and yield. In addition, the application of appropriate levels of inorganic fertilizers increases the productivity of land and labour and, at the same time, improves the efficiency of total energy use in agricultural production.

References

Agriculture and energy (1977) W. Lockeretz (Ed.) Academic Press, pp. 750.

Energy management and agriculture (1981) Royal Dublin Society, Ballsbridge, Ireland.

Green, N.B. (1978) Eating oil, energy use in food production. Westview Press, pp. 205.

Jürgens-Gschwind, S. (1980) Agriculture and energy with special reference to mineral fertilization. Fertilizer Research 1, 137–157.

Leach, G. (1976) Energy and food production. IPC Science and Technology Press, pp. 137.

Lewis, D.A. (1979) Energy in UK agriculture. J. Sci. Food Agric. 30, 449–457.

Stout, B.A. (1979) Energy for world agriculture. FAO Agriculture Series No. 7, pp. 286.

White, D.J. (1981) Energy in agriculture. The Fertiliser Society London, Proc. No. 203.

L Miscellaneous

L 1 International organizations concerned in the environmental aspects of fertilizer use

(a) United Nations Agencies

FAO (Food and Agriculture Organization of the United Nations), Rome

This is the primary UN Agency concerned with all aspects of fertilizer use.

An expert Consultation on "Effects of Intensive Fertilizer Use on the Human Environment" was convened in Rome, 25–28 January 1972, by the Soil Resources Development and Conservation Service of FAO Land and Water Development Division, and the report on this Consultation, published by the Swedish International Development Authority under arrangements with FAO, as Soils Bulletin No. 16, is a basic document for research in this field. A small booklet (published in English, French and German editions) containing a summary review of the Consultation, together with the conclusions and recommendations, was widely distributed by FAO at the United Nations Conference on the Human Environment, Stockholm, June 1972.

The fertilizer industry maintains a Fertilizer Industry Advisory Committee Liaison Office at FAO headquarters in Rome.

UNEP (United Nations Environment Programme), Nairobi

The establishment of this Agency was a consequence of the United Nations Conference on the Human Environment, held in Stockholm in June 1972.

Recommendations 21 a/iv and 22a of the Stockholm Conference advocated a strengthening of international activities (by Governments and FAO) for the exchange of information and cooperative research and technical assistance to developing countries to support national programmes on, *inter alia*, dose and timing of the application of fertilizers and their effects on soil productivity and the environment, and increased emphasis (by FAO) on control and re-cycling of wastes in agriculture, including assistance to national activities relating to recycling of crop residues, animal wastes and agro-industrial waste and use of municipal wastes as fertilizers.

The fertilizer industry maintains contact with UNEP through a correspondent in Nairobi (IPI Agricultural Mission to East and Central Africa).

UNESCO (United Nations Educational, Scientific and Cultural Organization), Paris

Project 9 of the UNESCO Man and the Biosphere Programme deals with ecological evaluation of the effects of the use of pesticides and fertilizers on land and water ecosystems.

ECE (United Nations Economic Commission for Europe), Geneva

The Economic Commission for Europe was established by the Economic and Social Council of the United Nations (ECOSOC) in 1947. Membership includes a number of Eastern European countries which are not members of FAO. At least three of its Committees (Agricultural Problems, Water Problems, Chemical Industry) are concerned with possible environmental effects of fertilizers.

In particular, the Committee on Agricultural Problems (in conjunction with FAO) held a Symposium in June 1974 on the effects of fertilizers on the quality and nutritional value of grains, potatoes, selected fruits and vegetables and forage. A Seminar, held at the IAEA headquarters in Vienna, on the pollution of waters by agriculture and forestry, was organized by the Committee on Water Problems in October 1973, at which Problem III was concerned with effects of the increased use of fertilizers on the pollution of water bodies (the contribution from FAO Soil Resources Development and Conservation Service to this session was particularly worthy of note). The Chemical Industry Committee's interest in environmental problems is concerned with fertilizer manufacture rather than fertilizer use.

UNIDO (United Nations Industrial Development Organization), Vienna

The first United Nations Interregional Seminar on the Production of Fertilizers was held in Kiev, Ukrainian SSR, in September 1965. The Second Interregional Fertilizer Symposium, organized by UNIDO in cooperation with the Governments of the USSR and India, was held in Kiev and New Delhi in September/October 1971 (report published under the title "Recent Developments in the Fertilizer Industry", United Nations, New York, 1972). One of the conclusions of the latter Symposium was that problems of environmental pollution arise during production in fertilizer plants and in the use of fertilizer products in agriculture; the associated recommendation 21 was that "UNIDO should proceed with the proposed "global project" for UNDP financing on the effect of the manufacture, distribution and use of chemical fertilizers on the environment, and the control of pollution". Close liaison is maintained by UNIDO with FAO on matters concerning fertilizers.

WHO (World Health Organization), Geneva

Publications of special interest include "European Standards for Drinking Water" (WHO Regional Office for Europe, Copenhagen: 1st ed. 1961, 2nd ed. 1970, 3rd ed. 1972), "International Standards for Drinking Water" (first published in 1958), and an article entitled "Food Facts" by J.C. Witschi and F.J. Stare of the Harvard School of Public Health, Boston, USA, which appeared in a booklet "Better Food for a Healthier World" for World Health

Day, 7 April 1974. The latter states categorically that "there is no difference in nutritive value between organically grown food and food grown with the aid of chemical fertilizers and chemical pesticides".

IAEA (International Atomic Energy Agency), Vienna

This agency has a special interest in isotope studies of agricultural problems. A booklet entitled "Effects of Agricultural Production on Nitrates in Food and Water with particular reference to Isotope Studies", published in 1974, comprises the proceedings and report of a panel of experts organized by the Joint FAO/IAEA Division of Atomic Energy in Food and Agriculture in June 1973.

(b)   European intergovernmental organizations

OECD (Organization for Economic Cooperation and Development), Paris

Fertilizers are of interest to the OECD Environment Directorate, to the OECD Development Centre and to the Committees on Agriculture and Industry. The fertilizer industry participated in 1971/72, in an observer capacity, in a Working Group on Fertilizers and Agricultural Waste Products, set up by the Water Management Sector Group, Programme on Evaluation of Eutrophication Control, whose report was published in 1973 by OECD Environment Directorate under the title "Impact of Fertilizers and Agricultural Waste Products on the Quality of Waters". This includes a technical report by G.J. Kolenbrander which is a valuable source of information on losses of N and P in leaching and run-off.

Council of Europe, Strasbourg

The *Consultative Assembly* Agriculture Committee has received two reports dealing with environmental aspects of fertilizer use, the first (January 1972) from Dr. Barry Commoner presenting a case which was heavily biased against fertilizers, and the second (January 1973) from Professor E. Welte presenting a more balanced view.

The Committee of Ministers, European Committee for the Conservation of Nature and Natural Resources, has published a European Soil Charter and a Study on Soil Preservation.

EEC Commission (Commission of the European Communities), Brussels

Having previously relied on FAO and OECD for information, the Commission has itself taken a direct interest in environmental aspects of fertilizers since 1971. Three divisions of the Commission are now concerned: Agriculture, Industrial and Technological Affairs and Environment.

A study undertaken for the Directorate-General of Agriculture by Professor A. Noirfalise of the Faculty of Agronomic Sciences at Gembloux has been published in French under the title "Conséquences écologiques de l'application des techniques modernes de production en agriculture" (Informations Internes sur l'Agriculture No. 137, November 1974).

A study by Dr. L. Whalley of the Warren Springs Laboratory, Stevenage, on pollution problems resulting from the manufacture of nitrogenous and phosphate fertilizers was commissioned by the Directorate-General of Industrial and Technological Affairs in 1973.

(c) Industry Associations

ICC (International Chamber of Commerce), Paris

This organization decided to establish an International Centre of Industry and the Environment, in Nairobi, as a link between industry and UNEP at the international level. The European fertilizer industry is represented through CEFIC (see below).

CEFIC (Conseil Européen des Féderations de l'Industrie Chimique), Brussels

This association of Chemical Industry Associations has set up a Working Group on Environmental Protection, three sub-groups of which have a limited interest in fertilizers: Surface and Ground Waters, Waste and Product Use.
   The NPK Working Group on Environmental Aspects of Fertilizer Use (see Introduction) is directly represented on the Products Use sub-group, and IFA — and, in theory, also APEA (see APEA) — on the main Working Group. Many of the individual members of APEA/CEA and IFA are represented on CEFIC through their national associations, and there is close liaison between APEA/ CEA/IFA and CEFIC in relations with the EEC Commission.

CEA (Centre d'Etude de l'Azote), Zurich

The aim of this international association of major Western European producers and distributors of nitrogenous fertilizers is "the scientific and practical study of methods capable of ensuring a rational and increasing use of nitrogenous fertilizers throughout the world". CEA, with its Section Agricole, is greatly concerned with all problems that may arise through fertilizer use. In the particular sphere of environmental problems, CEA works in close collaboration with the phosphatic and potassic fertilizer associations IFA and IPI, with its sister organization APEA, and with a wide range of international and regional organizations having an interest in this field.

APEA (Association des Producteurs Européens d'Azote), Zurich

This association of major Western European producers and distributors of nitrogenous fertilizers shares the same headquarters as CEA. As regards the environment, its main interest is in problems of pollution arising from fertilizer manufacture, and in legislation and control. It, too, works in close collaboration with IFA, particularly in relation to liaison with the EEC Commission and other international organizations.

IFA (International Fertilizer Industry Association), Paris

IFA is a worldwide association aimed to promote the production and use of all chemical fertilizers. Through its Environmental Sub-Committee which was set up by its Agricultural Committee in 1971, ISMA (now IFA) joined with CEA and IPI in the formation of the NPK Working Group on Environmental Aspects of Fertilizer Use. An Environmental Adviser was appointed in 1973, to ensure closer and more regular relations with governmental associations concerned with environmental problems. There is a joint IFA APEA Working Party on safety problems arising from fertilizer manufacture.

IPI (International Potash Insitute), Worblaufen-Berne

IPI's approach to environmental problems, with support from the agricultural research stations of its producer members in France and Germany, is mainly in the context of its documentation and publication services and of the annual scientific meetings (Congresses and Colloquia) planned by the Scientific Board.

(d) Other Non-governmental Associations

CIEC (Centre International des Engrais Chimiques), Zurich

This association organizes a series of World Fertilizer Congresses attended by scientists from Eastern European as well as from Western European countries.

CIRCA (Centre International pour la Coordination des Recherches en Agriculture), Zurich

A parallel association to CIEC, sharing the same address, CIRCA held a Symposium on "Intensive Agriculture and the Environment" at Newcastle-upon-Tyne in September 1973.

NGOs (Non-Governmental Organizations to official Conferences)

This loosely organized group is formed on the occasion of major world conferences, notably the United Nations Conference on the Human Environment in Stockholm and the World Population Conference in Bucharest, by representatives of non-governmental organizations covering a very wide range of interests, and expressing varying, and at times strongly divergent, opinions on the role of fertilizers and agricultural chemicals in world food production. This group has met on several occasions since the Stockholm Conference and has been given some encouragement by UNEP and constitutes a potentially important environmental lobby.

CAST (Council for Agricultural Science and Technology), Ames, Iowa, USA

CAST is a consortium of 25 national and regional food and agricultural science societies working to advance the understanding and use of food and agricultural science in the public interest. CAST seeks to improve the quality of national decision-making by preparing up-to-date reports on the scientific facts of food and agricultural issues. Multidisciplinary task forces of eminent scientists compile the reports as a service to the American Congress, the regulatory agencies, and the news media. They are also useful to researchers, students, opinion leaders and others interested in the state-of-the-science of selected critical issues. Subscriptions for CAST publications are available to libraries and other information centres. Each subscriber receives one copy of every publication issued within the calendar year.

L 2 Sources of information on environmental aspects of fertilizer manufacture

Specific information concerning environmental aspects of fertilizer manufacture can be obtained as follows:

Nitrogenous fertilizers:	Association des Producteurs Européens d'Azote (APEA) Bleicherweg 33 CH–8002 Zurich
Phosphatic fertilizers:	International Fertilizer Industry Association (IFA) 28, rue Marbeuf F–75008 Paris
Potassic fertilizers:	International Potash Institute (IPI) P.O. Box 41, Metrohaus CH–3048 Worblaufen-Bern
For questions concerning more general aspects of the chemical industry apply to:	Conseil Européen des Fédérations de l'Industrie Chimique (CEFIC) Working Party on Protection of the Environment Secretariat 250, Avenue Louise (Boîte 71) B–1050 Brussels